FORMATIVE ASSESSMENT AND SCIENCE EDUCATION

Science & Technology Education Library

VOLUME 12

SCOPE

The book series *Science & Technology Education Library* provides a publication forum for scholarship in science and technology education. It aims to publish innovative books which are at the forefront of the field. Monographs as well as collections of papers will be published.

The titles published in this series are listed at the end of this volume.

Formative Assessment and Science Education

by

BEVERLEY BELL

and

BRONWEN COWIE

University of Waikato, Hamilton, New Zealand

KLUWER ACADEMIC PUBLISHERS
DORDRECHT / BOSTON / LONDON

A C.I.P. Ctalogue record for this book is available from the Library of Congress.

ISBN 0-7923-6768-5 (HB)
ISBN 0-7923-6769-3 (PB)

Published by Kluwer Academic Publishers,
P.O. Box 17, 3300 AA Dordrecht, The Netherlands.

Sold and distributed in North, Central and South America
by Kluwer Academic Publishers,
101 Philip Drive, Norwell, MA 02061, U.S.A.

In all other countries, sold and distributed
by Kluwer Academic Publishers,
P.O. Box 322, 3300 AH Dordrecht, The Netherlands.

Printed on acid-free paper

Printed in the Netherlands.

CONTENTS

ACKNOWLEDGMENTS

We wish to acknowledge and thank the following people for their help and support in the writing of this book:

• The teachers and students involved in the Learning in Science Project (Assessment), who took the risk to have us in their classrooms, researching their assessment practices, and who talked so openly and honestly about assessment.

• The New Zealand Ministry of Education for funding the Learning in Science Project (Assessment).

• The University of Waikato, Hamilton, New Zealand for funding associated with the research.

• Those colleagues with whom we have discussed formative assessment, including Fred Biddulph, Paul Black, Carol Boulter, Sally Brown, Margaret Carr, Terry Crooks, Christine Harrison, Alister Jones, John Pryor, Phil Scott, Merilyn Taylor, Harry Torrance.

• The following publishers for permission to reprint material:
Chapter 5   Cowie, B. and Bell, B. (1999). A Model of Formative Assessment in Science Education, *Assessment in Education*, 6 (1) 101-116.

>Taylor and Francis Ltd
>PO Box 25, Abingdon
>Oxfordshire, OX14 3UE
>United Kingdom

Chapter 7.3   Bell, B. (in press) Formative assessment and science education; a model and theorising. In R. Millar, J. Leach, J. Osborne (Eds) (in press) *Improving Science Education: the Contribution of Research*. Buckingham, Open University Press.

>Open University Press
>Celtic Court
>22 Ballmoor
>Buckingham MK18 1XU
>United Kingdom

CHAPTER 1

INTRODUCTION

> It was the last lesson on Monday. The students had already told the teacher they had studied separating mixtures in previous years. The lesson started with a class discussion on filtering, decanting, crystallising and distilling. The teacher introduced these words by making links with the students' everyday experiences. They talked about 'decanting' cooked potatoes and compared filtering with sieving. Then she introduced a 'thinking exercise'. The students were to think about how they would separate a list of mixtures. The class discussed how to separate the first mixture which was broad beans and kidney beans. The next mixture was oil and water.
> The teacher moved around the class talking with the students. One group called her over to confirm they should filter the oil and water.
> The teacher talked with them and then went to the prep room, returning with oil, water and filter paper. She poured some oil on the filter paper, and mixed some oil and water. The students asked questions and made suggestions. The teacher moved towards the front of the class. As she did another group asked how big salt crystals were and told her that they would separate sand and salt using tweezers. The teacher stopped the class, showed them the oil and water mixture and said:
> If you are not sure for oil and filter paper ......". (Cowie, 1997)

This can be considered as an instance of formative assessment as it involved the teacher gathering, interpreting and acting on information about the students' learning, in order to improve the learning, during the learning. This episode is one of many which were observed during the research (Bell and Cowie, 1997) documented in this book.

Formative assessment is increasingly becoming a focus in policy documents on educational assessment and in the professional development of teachers. The term 'formative assessment' is not new but is now being used in more detailed and specific ways. And as this happens, there is a call for further research and theorising on formative assessment (Black and Wiliam, 1998). This book documents research that explored the current practice of some science teachers to describe, explain and theorise about the formative assessment being done in some New Zealand science classrooms.

But first some background to the term 'formative assessment'. There have been three trends in education that have highlighted the need for teachers to do formative assessment.

1.1 CONTINUOUS SUMMATIVE ASSESSMENT

One of the trends in educational assessment that has put the spotlight on formative assessment, is the development of more valid assessment procedures. In the 1970s and 80s, there was much criticism of the validity of summative assessments used in educational assessment, and in particular, of the limitations of the validity of external testing and examinations (Keeves and Alagumalai, 1998). This included criticisms of the validity of assessment tasks such as multiple choice questions and criticisms of norm-referenced assessments such as those for national qualifications. There was also criticism of the impact of high stakes, standardised testing on school learning (Black and Wiliam, 1998). The responses to these criticisms can be summarised as a need:

• to assess a wider range of science learning outcomes, such as performance of investigation skills (Johnson, 1989) and multiple forms of intelligence (Gardner, 1985);

• to use a wider range of assessment tasks (other than multiple choice tests, questions requiring short answers and essay questions), for example, portfolios (Gitomer and Duschl, 1995, 1998; Duschl and Gitomer, 1997); and performance based assessment (Erickson and Meyer, 1998).

• to integrate assessment with the curriculum and to assess in more authentic contexts (Tamir, 1998)

As these recommendations could not be achieved through external examinations or standardized testing alone, assessment by teachers (also called internal assessment) was seen as a way forward. Hence, an early use of the term 'formative assessment' was to distinguish between continuous summative assessment by teachers in the classroom and summative assessment by external examiners, such those who develop standardized tests and those who set and mark examinations for national qualifications. This continuous summative assessment by teachers was initially called formative assessment as it did enable some information on learning to be given to students and teachers in the course of the school year, although it was relatively coarse feedback. This has been called 'weak formative assessment' (Brown, 1996). The questions often raised during discussions on continuous summative assessment were those such as how many separate assessments have to be recorded for the aggregated mark or grade to be reliable and valid; how best to store the multiple assessment documentation; how to aggregate the marks or grades; the problems with reducing many assessment results into one grade; and whether all the achievement objectives in the science curriculum have to be assessed and how often.

1.2 MULTIPLE PURPOSES FOR ASSESSMENT

Another trend, which has highlighted formative assessment, was the trend towards multiple purposes for assessment. This trend was brought into sharp focus in the 1990s, when politicians, and others wanting to hold educationalists accountable, looked to assessment to provide the information required for the accountability process. This added to the existing demands for assessment information by people who operate outside the classroom, for example, care-givers, principals, school governing bodies, local or national government officials, awarders of national qualifications, selection panels for tertiary education programmes, and employers. In New Zealand and elsewhere internationally, this trend towards using educational assessments for accountability purposes in addition to the existing purposes, has highlighted the multiple purposes for assessment. These multiple purposes can include auditing of schools, national monitoring, school leaver documentation, awarding of national qualifications, appraisal of teachers, curriculum evaluation and the improvement of teaching and learning.

An indication of this trend in New Zealand was evident in the late 1980s. In 1989, the government issued a discussion paper for national consultation called

'Assessment for Better Learning' (Department of Education, 1989). Submissions were invited from professionals in education and the public on a range of assessment issues related to accountability. The terms of reference included:

'The working party shall:
1. recommend to the Government procedures for assessment which:
i. can monitor the effectiveness of the New Zealand school system on student learning;
ii. assess the effect of individual schools on students' learning achievements;
2. recommend ways of reporting on the above, taking into account different audience needs;
3. within the context of New Zealand's dual cultural heritage, advise the Government on the possible effects of such assessment and reporting procedures for students, teachers, the curriculum, schools, employers, and the wider community; ..'
(Department of Education, 1989, p5)

The report summarising the submissions to the ministerial working party on 'Assessment for Better Learning' was titled 'Tomorrow's Standards' (Ministry of Education, 1990). These two policy documents reflect the three cornerstones of the accountability process – a prescribed set of standards, an auditing and monitoring process to ascertain if the standards have been attained, and a way of raising standards if low standards have been indicated in the audits. In New Zealand, the 'standards' are contained in the New Zealand Curriculum Framework and associated documents, including those of the New Zealand Qualifications Authority (Ministry of Education, 1993a); the auditing is done by the Education Review Office and the monitoring by the National Education Monitoring Project.

The 'raising of standards' was seen by policymakers in New Zealand as been achievable by a number of methods, including school-based assessment (Ministry of Education, 1993a). It is school-based assessment that is the focus of the research reported in this book.

There is now a recognised need to clarify these multiple purposes for assessment. In a recent example, in the USA, a working party (National Research Council, 1999) identified three purposes of assessment:

'Assessment has multiple purposes. One purpose is to monitor educational progress or improvement. Educators, policymakers, parents and the public want to know how much students are learning compared to the standards of performance or to their peers. This purpose, often called *summative assessment*, is becoming more significant as states and school districts invest more resources in educational reform.
A second purpose is to provide teachers and students with feedback. The teachers can use the feedback to revise their classroom practices, and the students can use the feedback to monitor their own learning. This purpose, often called *formative assessment*, is also receiving greater attention with the spread of new teaching methods.
A third purpose of assessment is to drive changes in practice and policy by holding people accountable for achieving the desired reforms. This purpose, called *accountability assessment*, is very much in the forefront as states and school districts design systems that attach strong incentives and sanctions to performance on state and local assessments.' (National Research Council, 1999, pp1-2)

In New Zealand, the multiple purposes for assessment are acknowledged in the Curriculum Framework (Ministry of Education, 1993a):

'Assessment in the New Zealand Curriculum is carried out for a number of purposes. The primary purpose of school-based assessment is to improve students' learning and the quality of learning programmes. Other purposes of assessment include providing feedback to parents and students, awarding qualifications at the senior level, and monitoring overall national educational standards. Assessment also identifies learning needs so that resources can be effectively targeted. To meet these different purposes, a range of assessment procedures is required.' (p24).

In New Zealand, these multiple assessment purposes were seen as being addressed by using the multiple procedures of school-based assessment. The purposes of school-based assessment are described as 'improving learning, reporting progress, providing summative information, and improving programmes' (Ministry of Education, 1994, p. 7-8). These multiple purposes of school-based assessment are seen as giving rise to three broad categories of assessment: diagnostic, summative and formative assessment.

Formative assessment is described in the policy document as:

'Formative assessment is an integral part of the teaching and learning process. It is used to provide the student with feedback to enhance learning and to help the teacher understand students' learning. It helps build a picture of a students' progress, and informs decisions about the next steps in teaching and learning.' (Ministry of Education, 1994, p. 8).

While the above description of formative assessment could include continuous summative assessment, the research documented in this book specifically explores formative assessment as classroom assessment to improve learning (and teaching) during the learning.

Multiple purposes for assessment means that there are multiple audiences, and it raises the issue of whether one assessment task can provide information for several assessment purposes and audiences (Black,1998). There is a need to re-evaluate the appropriateness of procedures to meet these goals for specified audiences. There is also a need to prioritise the different purposes. For example, the review by Black and Wiliam (1998) asserts that the context of national or local requirements for certification and accountability will exert a powerful influence on the practice of assessment, to the detriment of formative assessment (p. 20). In New Zealand, there is a growing recognition that school-based assessment needs to emphasise assessment for improving learning, more than assessment for accountability (Hill,1999).

1.3 TEACHING FOR AND ASSESSMENT OF CONCEPTUAL DEVELOPMENT

The third trend that has highlighted formative assessment is the development of views of assessment to match the views of learning, which recognise that each learner has to construct an understanding for her or himself, using both incoming stimuli and existing knowledge, and not merely absorbing transmitted knowledge (Gipps, 1994; Berlak, 1992a,b; Wiliam, 1994). These views of learning acknowledge that both students' existing knowledge and thinking processes influence the learning outcomes achieved and therefore both need to be taken into account in teaching (Bell, 1993a). In taking into account students' thinking in their teaching, teachers are responding to and interacting with the students' thinking that they have elicited in the classroom. They are therefore undertaking formative assessment whilst teaching for conceptual development.

In science education, teaching for conceptual development arose from the 1980's research on children's alternative conceptions (Driver, 1989). A central part of this teaching is dialogue (not a monologue) with students to clarify their existing ideas and to help them construct the scientifically accepted ideas (Scott, 1999). Therefore, giving feedback to students about how their existing conceptions relate to the scientifically accepted ones, and helping them to modify their thinking accordingly, is both a part of formative assessment and teaching for conceptual development. Formative assessment is seen as a crucial component in teaching for conceptual development (Bell, 1995). Consider this scenario (adapted from Tasker and Osborne, 1985):

> The students are busy in a lesson doing an investigation into electrolysis of copper chloride. The students are working in pairs and are following the instructions on the whiteboard and which they had discussed with the teacher at the beginning of the lesson. The teacher is moving between each group, listening to the students talking and being available to answer questions if need be. She is stopped by a group who are concerned that bubbles of gas are coming off one electrode only and not both. They ask the teacher what is wrong with their circuit as one electrode seems not to be working. The teacher spends time with the group to elicit their view of electric current. All three students in the group think that the current comes from both electrodes and clashes in the solution of copper chloride. They assume that gas would be formed at each electrode if all was working well. The teacher interprets the students as having an alternative conception about electric circuits (the clashing currents model) and decides to spend the second half of the day's lesson addressing this with the whole class.

This is an instance of formative assessment. The teacher created opportunities (using the group work situation) to interact individually or with small groups of students. She noticed and recognised that some of the students held scientifically unacceptable ideas and responded to address the problem with some teaching later in the lesson. It is also an example of teaching for conceptual development, which takes into account students' existing thinking.

Research into teacher development to help teachers improve their practice to take into account students' thinking, indicated that teachers are asking for more information and professional development on how to elicit, respond to and interact with students' thinking and knowledge in the classroom (Bell, 1993b). For example, they are seeking more assistance to be able to respond to the student who says 'there is no gravity on the moon because there is no atmosphere'. Responding to this elicited information requires the teacher to identify that this view is not the scientifically accepted one and that it is a commonly held alternative conception by students of all ages. The teacher is also required to act in some way so the student has opportunities and help to learn the scientifically accepted concept of 'gravity'. This will involve providing learning opportunities for the students to explore their own ideas of gravity, to have presented to them the scientific view of gravity, to modify their own ideas, and to use the new ideas with confidence (Driver, 1989). Formative assessment is a key part of teaching which takes into account students' thinking.

Likewise, research has also indicated that the greatest source of feedback to teachers, when they are changing their teaching, is feedback that the new teaching activity is resulting in 'better learning' (Bell and Pearson, 1992). 'Better learning' was seen by the teachers in the research as both 'better learning' conditions and 'better learning' outcomes. 'Better learning' conditions included increased enjoyment, social co-operation, ownership, student confidence, and motivation. 'Better learning' outcomes included the responses to teacher questions, debates and written work, conceptual development and transfer of learning. The teachers in the research tended to use 'better learning conditions' more than 'better learning outcomes' as a source of feedback on the effectiveness of their teaching. The teachers requested further professional help on how to elicit and act on information about student learning outcomes, while the teaching and learning is occurring in their classrooms, so as to improve their own teaching as well as the students' learning.

1.4 FORMATIVE ASSESSMENT AND LEARNING

Due to these three trends in education, formative assessment is increasingly being used to refer only to assessment which provides feedback to students (and teachers) about the learning which is occurring, during the teaching and learning, and not after. The feedback or dialogue is seen as an essential component of formative assessment interaction where the intention is to support learning (Clarke, 1995; Sadler, 1989; Perrenoud, 1998).

And assessment can be considered formative only if it results in action by the teacher and students to enhance student learning (Black, 1993), for example:

> The distinguishing characteristic of formative assessment is that the assessment information is used, by the teacher and pupils, to modify their work in order to make it more effective. (Black, 1995a)

> Formative assessment has been defined as the process of appraising, judging or evaluating students' work or performance and using this to shape and improve students' competence. (Gipps, 1994)

It is through the teacher-student interactions during learning activities (Newman, Griffin and Cole, 1989) that formative assessment is done and that students receive feedback on what they know, understand and can do. It is also in these student-teacher interactions during learning activities that teachers and students are able to generate opportunities for furthering the students' understanding. As formative assessment is viewed as occurring within the interaction between the teacher and student(s), it is at the intersection of teaching and learning (Gipps, 1994, p. 16). In this way, teaching, learning and assessment are integrated in the curriculum.

Therefore, the process of formative assessment always includes students. It is a process through which they find out about their learning. The process involves them in recognising, evaluating and reacting to their own and / or others' evaluations of their learning. Students can reflect on their own learning or they may receive feedback from their peers or the teacher.

Formative assessment is also the component of teaching in which teachers find out about the effectiveness of the learning activities they are providing. It can be viewed as the process by which teachers gather assessment information about the students' learning and then respond to promote further learning. For example:

> 'Assessment should contribute to instruction and learning. ..Assessment after instruction is over does not allow for the assessment to contribute to any instructional decisions. All that can be said is the degree to which a student mastered some amount

of content. Assessment must be a continuous process that facilitates "on-line" instructional decision making in the classroom' (Gitomer and Duschl, 1995, p. 307)

Both formative and summative assessment influence learning. In other words, to improve learning outcomes, we need to consider not only the teaching and learning activities but also the assessment tasks. Gipps and James (1998) summarised the ways in which assessment influences learning in four main ways. Firstly, assessment can provide a motivation to learn by giving a sense of success in the subject (or demotivation through failure) and through giving a sense of self-confidence as a learner. Secondly, assessment can help students (and teachers) decide what to learn by highlighting what is important to learn (it may not be necessary to learn all that is taught) and by providing feedback on success so far. Thirdly, assessment helps students learn how to learn by encouraging an active or passive learning style; by influencing the choice of learning strategies; by inculcating self-monitoring skills; and by developing the ability to retain and apply knowledge and skills and understanding in different contexts. Lastly, assessment helps students learn to judge the effectiveness of their learning by evaluating existing learning; by consolidating or transforming existing learning and by reinforcing new learning.

Moreover, the extent to which formative assessment improves learning outcomes is now being recognised. For example, Black and Wiliam (1998) in their review of classroom assessment boldly state:

The research reported here shows conclusively that formative assessment does improve learning. The gains in achievement appear to be quite considerable, and as noted earlier, amongst the largest ever reported for educational interventions (p. 61).

While there has been much written on the importance of formative assessment to improve learning and standards of achievement (Harlen and James, 1996), there has been little research on the process of formative assessment itself. And, as Black and Wiliam (1998) suggest, there is a need to explore views of learning and their inter-relationships with assessment.

1.5 THE LEARNING IN SCIENCE PROJECT (ASSESSMENT)

This book reports on the findings of a research project investigating formative assessment in some science classrooms in New Zealand. This research – the Learning in Science Project (Assessment) – is fully documented in Bell and Cowie (1997). This research was done under contract to the New Zealand Ministry of Education in 1995-96 to investigate classroom-based assessment in science education in Years 7-10 (ages 11-14 years) classrooms where the teacher of science was taking into account students' thinking (Bell, 1993a). Four key aims for the research were:

1. to investigate the nature and purpose of the assessment activities in some science classrooms.

2. to investigate the use of the assessment information by the teacher and the students to improve the students' learning in science.

3. to investigate the teacher development of teachers with respect to classroom-based assessment, including formative assessment.

4. to develop a model to describe and explain the nature of the formative assessment process in science education.

Formative assessment in this research was defined as:

> the process used by teachers and students to recognise and respond to student learning
> in order to enhance that learning, during the learning (Cowie and Bell, 1996).

The focus of this research was on formative assessment and not on assessment for qualifications, reporting to parents and care-givers, school-leaver documentation or assessment for inspection or audit agencies. While these latter aspects of assessment in primary and secondary education cannot be separated off entirely, they were not the focus of this research. And while continuous summative assessment was seen as formative assessment in the past, in this research a distinction was made between the terms 'formative assessment' and 'continuous summative assessment'. Continuous summative assessment is a term used to describe the continuous assessment of student learning, which is recorded over an extended period of time, aggregated and reported to the student and others at some later date. This accumulation of assessment information was often referred to as formative assessment in the past. By the definition of formative assessment used in this research, it would only be considered as formative assessment if some action to improve learning during the learning was involved. As this is usually not the case with continuous summative assessment, it has not been considered as formative assessment in this research.

The research was mainly qualitative, interpretive, collaborative and guided by the ethics of care. Multiple data collection techniques were used, including interviews, surveys, and participant observation. A fuller documentation of the research methodology is given in Bell and Cowie (1997, 1999).

The research had three strands:

• Ideas about assessment

In this strand of the research, the views of assessment of nine teachers of science and one teacher of technology and some of their students were elicited at the beginning of the project and monitored throughout the project. Data for this strand were collected through interviews and surveys.

• Classroom-based studies

In this strand of the research, the classroom assessment activities, and in particular the formative assessment activities, of the ten teachers and their students were studied and documented in the form of eight case studies (the data generated by teachers 1, 4, 6 were written as one case study). Case studies were chosen as one level of analysis to investigate the multiple and integrated social and cognitive processes involved in formative assessment. Data for this strand were collected by participant observation, involving field notes, head notes, and documentary data such as the writing on the board, student books, the wall displays, the teachers' plan' for the unit and the teachers' record books.

• Teacher Development Studies

The research intentionally combined research and development activities (Bell and Cowie, 1999), with one strand of the research being a teacher development strand (Bell and Cowie, 1997, pp. 259-277). It was felt necessary to include a developmental strand in the project for four reasons. Firstly, the researchers held a view that the research process should have reciprocal purposes and gains for both the teachers and researchers. However, the gains for the researchers and teachers may not be the same. Whereas the main aim for the researchers was the creation of new knowledge about classroom-based assessment, teachers in previous research projects had indicated that they often got involved in major research projects for the opportunities for professional development. The teachers valued these opportunities for sharing ideas with other teachers, time for reflection, the input of new theoretical ideas and classroom activities, the support for trialing new classroom activities and for the information about wider educational developments (Bell and Gilbert, 1996). These activities could best be fostered in the teacher development days although it is also acknowledged that they also occurred in the data collection activities of interviews, surveys and classroom observations.

Secondly, the researchers felt that the teachers did not necessarily have the awareness and language to discuss the phenomenon being researched, that is, formative assessment. It was felt that some professional development activities would enable the teachers to develop their skills of and knowledge and language about formative assessment so that they could discuss it in a way that would aid the data collection and analysis for the research. The interviews also aided in this.

Thirdly, the teacher development days were included so that the teachers and researchers could meet to discuss the emerging data analysis. The discussions provided a secondary data generation and collection opportunity for the researchers and further reflective opportunities for the professional, personal and social development (Bell and Gilbert, 1996) of the teachers.

Lastly, data to inform future teacher development courses on classroom-based assessment was sought.

In summary, the research was investigating the existing assessment practice of the ten teachers. But on the other hand, it was investigating their developing assessment practices over the two years of being involved in a research project. In this strand of the research, teacher development activities were undertaken by the ten teachers to develop the formative assessment activities they used in their classrooms and to reflect on the data collected and analysed. This occurred on eleven days over 1995-1996 and was facilitated by the researchers. Data for this strand were collected by audiotaped discussions, surveys and field notes.

Timewise, the research was also divided into two parts. The first phase was undertaken in January - June, 1995 (Cowie and Bell, 1995) and researched the teachers and students views on assessment at the beginning of the research. The second phase started in July, 1995. From July to December 1995, classroom observations and interviews with five teachers and their students were undertaken. During this part, the framework for the data collection during the classroom observations for all the case studies was generated. The remaining five case studies were completed during January - October, 1996 and the development of the model of formative assessment was undertaken.

The ten teachers , who volunteered to take part in the research, were primary (middle school) and secondary (junior high) teachers, women and men, beginning and experienced teachers, and some had management responsibilities in the school. Some of the teachers had had previously experiences of working in a research project in science education. Where possible, the teachers came in pairs from each school involved so that they would have a buddy to discuss the research with.

Each of the ten teachers chose a class to work with them on the project. For each teacher, this class of students changed in January, 1996, at the start of the new school year. In total, there were 114 student interviews done during the course of the research.

The data documented in this book are illustrative rather than representative, given the constraints on space. Readers are referred to Bell and Cowie (1997) for a fuller documentation of the research findings and data. The coding of the data is explained in the appendix. In choosing quotations, no judgement was made as to whether the transcript was indicative of 'good' teaching practice or not. The quotations were chosen on the basis of the way in which they were illustrative of formative assessment.

While the research was conducted in the context of science education, the findings and discussion of them in this book will be of interest to educators working in other subject areas. We therefore, at times, discuss formative assessment in general – it being very clear that these insights were derived from the context of science education alone.

1.6 THE THEME OF THE BOOK

The theme of the book and our main argument is that formative assessment can be viewed as a purposeful, intentional activity; an integral part of teaching and learning; a situated and contextualised activity; a partnership between teacher and students; and involving the use of language to communicate meaning. We argue that formative assessment can be theorised as a sociocultural and discursive activity, and hence linked to sociocultural and discursive views of learning.

The book aims to describe and explain the formative assessment activities currently being used by the teachers involved in the research. It is hoped that this documentation will aid other teachers, preservice teachers, teacher educators and researchers to understand what formative assessment is and to further develop their skills and knowledge of formative assessment. The book also aims to theorise about formative assessment, not in relation to the accountability literature but with respect to the literature on learning.

Chapter 2 reviews the relevant literature on formative assessment, while chapter 3 documents a case study of the formative assessment in one teacher's classroom over the time of a unit of science work was being taught. In the next two chapters, two summaries of the data across all the case studies are given, first in the form of the characteristics of formative assessment (chapter 4) and secondly in a model of formative assessment (chapter 5). Further examples of formative assessment are given in chapter 6. In chapter 7, we theorise about formative assessment with respect to current theoretical views of learning. In the final chapter, we summarise the findings on teacher development to improve the practice of formative assessment.

CHAPTER 2

A REVIEW OF THE RELEVANT LITERATURE

This chapter briefly reviews the current literature on formative assessment. As outlined in the previous chapter, assessment which is intended to enhance teaching and learning is called formative assessment. Formative assessment is not a common term (Black and Wiliam, 1998) but as indicated in the previous chapter, it is increasingly being used in specific ways. The research documented in this book used the following definition of formative assessment:

> Formative assessment is defined as the process used by teachers and students to recognise and respond to students learning in order to enhance that learning, during the learning. (Cowie and Bell, 1996)

Other definitions include:

> Formative assessment has been defined as the process of appraising, judging or evaluating students' work or performance and using this to shape and improve students' competence. (Gipps, 1994)

> The distinguishing characteristic of formative assessment is that the assessment information is used, by the teacher and pupils, to modify their work in order to make it more effective. (Black, 1995a)

Within this research, diagnostic assessment, which aims to diagnose student weaknesses, is viewed as a part of formative assessment (Black, 1993). All these definitions emphasise that assessment is formative in its function only when action is taken which is intended to improve student learning; that formative assessment includes both the involvement of both teachers and students; and that formative assessment includes the notion of feedback, that is, 'any information that is provided to the performer of any action about that performance' (Black and Wiliam, 1998, p. 53). Formative assessment is that which supports learning.

Assessment to support learning is also referred to as educational or educative assessment. Gipps (1994) identified Glaser (1963) as the first to propose attention be given to 'educational assessment'. She argued that a shift from assessment to prove learning to assessment to improve learning was signalled by Glaser's suggestion of a move from norm to criterion referenced assessment. Educative assessment was also described as integral to learning as a process of feedback and as a dialogue between those actively participating in the learning environment (Blackmore, 1988; Willis, 1994).

In the next sections, formative assessment is discussed with respect to:

2.1 the process of formative assessment
2.2 feedback and formative assessment
2.3 formative assessment and its relationship with learning and teaching
2.4   the role of students and teachers
2.5 the impact of formative assessment

2.1 THE PROCESS OF FORMATIVE ASSESSMENT

Formative assessment is dependent on teachers' and students' mutual engagement in a process which involves them in eliciting, interpreting and acting on assessment information. The process is cyclic, all the aspects interact and are interdependent. The eliciting of information is discussed first.

Eliciting information

When the intention of assessment is to improve learning, diverse information needs to be gathered on how and what students are learning (Willis, 1994). The strategies used need to be able to gather information on the outcomes of student learning, as well as to gather the transient and ephemeral information which is produced during the process of learning. As different students are prepared and able to display their understandings in different ways, different modes for gathering information are required (Crooks, 1988; Stiggins, 1991).

Teachers gather a large amount of diverse information on student learning during informal interactions with them (Ministry of Education, 1994). They do this while observing, listening to and questioning students during whole class, small group and individual discussions and practical work. They also gather information by looking at written work. Suggestions to enhance the quality of information elicited this way include the use of open questions, probing and the use of increased 'wait times' after asking questions (Rowe, 1987). Much of the information which is gathered through informal interactions is not recognised by teachers or students as having a potentially formative function (Harlen, 1995). Some writers suggest that comprehensive and useful assessment information needs to be gathered more systematically (Black, 1995a; Harlen, 1995; Sutton, 1995). For example, Sutton (1995) suggested that teachers gather information during informal interactions over three week cycles and that during the fourth week they explicitly target those students about whom they know little.

Recent research has focused on the development of new strategies to gather assessment information and most of these new strategies foster student involvement in the process of assessment. They aim to stimulate students to display their thinking in a manner which serves as a focus for communication between teachers and students (Black, 1993). Such strategies include, for example, concept maps, portfolios, peer assessment (Sutton, 1995, White and Gunstone, 1992; Fairbrother, Black and Gill, 1995).

Another way of eliciting formative assessment information is self-assessment, which can be seen as a way to involve students in the assessment process itself (Parkin and Richards, 1995). Increased student involvement in assessment is consistent with constructivist views of learning (Bell and Gilbert, 1996, p. 44) which emphasise student thinking and metacognition. Not involving the students means there is only one perspective on any situation, and opportunities to clarify interpretations and to generate suggestions are lost. Teachers have also found that students' commitment to learning was strengthened when the students took more responsibility, in collaboration with the teacher, for monitoring their own progress, evaluating their own strengths and weakness and devising strategies for improving their learning (Fairbrother, Black and Gill, 1995; Klenowski, 1995). These researchers have found that an interactive dialogue between the teacher and the students during the self-assessment process was important in order to ensure a full analysis by the students. It is essential for students to have a clear overview of what they are expected to learn and the criteria used to judge whether the learning had been achieved, if they are to undertake self assessment and effectively focused action (Black, 1995a; Klenowski, 1995; Harlen and James, 1996; Boud, 1995; Clarke, 1995). It is only when a learner assumes ownership of and values a learning outcome that it can play a significant part in their voluntary self monitoring (Sadler, 1989; Raven, 1992):

> The indispensable conditions for improvement are that the *student* comes to hold a concept of quality roughly similar to that held by the teacher, is able to monitor continuously the quality of what is being produced *during the act of production itself*, and has a repertoire of alternative moves or strategies form which to draw at any given point. (Sadler, 1989; page 3)

Interpreting the information

The second part of the formative assessment process is that of interpreting the information. This involves making judgements, and the criteria used in interpreting formative assessment are an important consideration. Some writers argue formative assessment needs to be criterion-referenced and student-referenced if it is to be able to provide teachers with the information they need help students improve their learning (Black, 1995a; Harlen, 1995).

In criterion-referenced formative assessment, students' understandings are compared with a pre-determined set of criteria or descriptors which describe the levels at which it is possible to perform or achieve an outcome. The criteria locate what a student knows, understands or can do in relation to the desired learning outcome. Criterion-referencing strengthens the links between teaching and assessment (Black, 1993). However, the exclusive use of pre-determined criteria does not necessarily strengthen the links between assessment and what is learned because it assumes that students learn only what is taught (Biggs, 1995). It can fail to acknowledge the range of other learning and understandings which occur as the result of learning experience.

Some writers argue that in order to respond to student learning in a manner which recognises and optimises it, student-referenced assessment is also essential (Harlen and James, 1996). Student-referencing compares a student's learning with their own prior learning. However, student-referencing alone may not provide the information required to enable the teacher to provide effective feedback to guide student learning. The two forms of referencing interact to do this.

'Convergent' and 'divergent' assessment have been identified as two approaches to assessment used by teachers and which have links with student-referencing and criterion-referencing (Torrance and Pryor, 1995). Convergent assessment is defined as assessment which focuses on finding out if a student knows a predetermined thing. It is associated with detailed planning, the systematic collection of data and the 'interpretation of the interaction of the child and the curriculum from the point of view of the curriculum' (Torrance and Pryor, 1995). The implications of convergent assessment are essentially behaviourist. Divergent assessment is defined as finding out what students understand and is characterised by less detailed planning; open forms of recording; and an analysis of the interaction of the child and the curriculum from the point of view of the child. Divergent assessment supports a constructivist view of learning (Torrance and Pryor, 1995).

Torrance and Pryor's notion of divergent assessment shares some similarities with Sadler's (1989) notion of qualitative judgement and Wiliams' (1992) notion of construct-referenced interpretations. Sadler (1989) stated that qualitative judgements are required when learning is considered multi-dimensional as are holistic judgements of the quality of learning. He described qualitative judgements as holistic, invoking fuzzy criteria which were context dependent rather than predetermined:

> imperfectly differentiated criteria are compounded as a kind of gestalt and projected
> onto a single scale of quality, not by means of a formal rule but through the integrative
> powers of the assessor's brain. (Sadler, 1989, p. 132).

Hence, the salience of particular criteria depends on 'what is deemed to be worth noticing' at a particular time. To use qualitative judgements, teachers need to possess the concept of quality appropriate for the task and be able to judge students' work in relation to this. Qualitative judgements are important when assessing students' understanding of science concepts and for evaluating open-ended investigations or work which requires an extended response. Wiliam (1992) claimed that construct-referenced interpretations are required when assessing holistic and open-ended activities such as investigations, projects, creative writing and art works. Quality within these activities is not able to be completely encapsulated by a set of pre-determined criteria for 'the whole is more than the sum of the parts'. In both of these notions of qualitative judgement and construct-referenced interpretations, the criteria are emergent. That is, the criteria emerge or develop during the process of assessment. Judging quality requires connoisseurship and consensus is achieved through negotiation. Claxton (1995) argued that true self-assessment:

> must be embedded within a context that appreciates the intrinsic unspecificability of
> quality, and sees the discussion of criteria as a stage on the road towards developing
> an ability that is essentially intuitive.

Student-referencing, when interpreting information elicited in formative assessment, also enables teachers to take into account students' learning approaches, effort and progress, and their particular circumstances. It enables teachers to provide feedback which is sensitive to and supportive of the students as learners. For example, there may be a difference between the feedback given by a teacher to a learner who perceives the gap between her or his understanding and the desired understanding as so large that the goal of understanding seems unattainable, and the feedback given to another student who may view it as so small that closing it is not worth the effort.

Student-referencing enables teachers to support students' persistence, motivation and productive attributions (Harlen, 1995; Tunstall and Gipps, 1995). Crooks (1988) noted that assessment impacts on students' motivation and self-perception. Within constructivist views of learning (Bell and Gilbert, 1996, p. 44), motivation and persistence are relevant because learning is viewed as a process requiring conscious and deliberate activity. Many writers have noted that cognitive development cannot be separated from affective, motivational and contextual aspects (Berlak, 1992b; Hargreaves, 1989). Current theories of students' achievement motivation relate to the goals they have (Dweck, 1986, 1989), the attributions they make (Weinstein, 1989) and the impact of rewards (Lepper and Hodell, 1985). Certain goals and attributions produce more effective, sustained long term learning. Feedback can influence students' motivation and attributions. For example, feedback which conveys the message that mistakes are bad has been found to encourage competition rather than co-operation and to reinforce performance goals. Students with performance goals strive to successfully complete a task rather than to understand the underlying concepts (Dweck, 1986). Grades can lead students to attribute failure to low ability and reduce their motivation, confidence and effort (Black, 1993; Pryor and Torrance, 1996).

In summary, a combination of student-referencing and criterion-referencing enables teachers to interpret formative assessment information to enhance a broad range of learning outcomes.

Acting on the information

The third part of formative assessment is acting on the interpreted information to inform and improve student learning. Both teachers and students may take action but it is the student's action which is critical as they do the learning. Promoting effective action is not easy. Even when teachers have made distinctions between their students, they may not provide them with differentiated learning experiences (Bennett et al, 1984). Teachers may also have difficulty taking effective action when formative assessment reveals a diversity of individual understandings (Torrance, 1993; Dassa, Vazquez-Abad and Ajar, 1993; Black, 1993; Bennet et al, 1984; Savage and Desforges, 1995). Newman, Griffin and Cole (1989) raised questions about teachers' ability to identify individual achievement when attention is usually focused on groups or the whole class and when, as they claim, cognitive change is as much social as it is individual. Much teacher action appears to be directed towards the progress of the class through the curriculum, rather than individual development (Bachor and Anderson, 1994; Savage and Desforges, 1995)

Teacher actions may be planned for, as well as on-the-spot interventions. Planned actions can occur at a class, small group or individual level. When teachers have judged students as having different understandings, they have been observed to provide students with different tasks and / or materials to work with (Sutton, 1995; Savage and Desforges, 1995). They have also been observed to provide open-ended tasks which permit differentiation by outcome (Sutton, 1995). Teachers have been observed to hold an assessment part way through a topic and to provide differentiated experiences for students after this (Black, 1995a). Sutton (1995) argued that if teachers have made no provision for differentiated learning experiences, then the demands on them during a lesson are very high, and this can result in some groups of students going unnoticed. Planning for formative assessment and differentiated learning experiences is essential and must allow for flexibility within the learning programme so that teachers can respond to what they find out (Black, 1995a; Harlen and James, 1996). Perrenoud (1991) stated 'There is a desire in everyone not to know about things about which one can do nothing'. Black (1995a) also suggested the need for and difficulty of flexible planning may be one reason why formative assessment is poorly developed.

Teachers can also act on formative assessment information while interacting with students. Such actions or interventions are essentially spontaneous and provide students with on-the-spot feedback. Through on-the-spot actions, these teachers can take action at different times. They may choose to act immediately or to delay their action. Providing more practice, moving on to the next topic and re-explaining the topic were common actions in this kind of formative assessment (Bennett et al, 1984; Gipps, 1994). Savage and Desforges (1995) categorised this form of intervention as providing feedback in the form of general encouragement and social support; questioning; and further information. Tunstall and Gipps (1995) described teachers working with six and seven year olds as acting to provide social, evaluative and descriptive feedback. They described one form of descriptive feedback as essentially behaviourist in that it identified errors in relation to the teachers' goals. They described another form as deriving from a constructivist view in that it was interactive and used both pre-determined and emergent criteria. This form of descriptive feedback provided feedback in the form of praise and included the generation of possibilities for student action. Teachers have been observed to defer action until a later stage (Harlen, 1995). Some teachers deferred their action until they had more comprehensive, detailed and actionable information or until they considered their action would be more effective. Teachers also used their knowledge of the contexts in which their students could or could not do something to provide them with additional information and to help them take more effective action (Harlen, 1995).

The use of student self-assessment is a planned action which has been found to be effective in encouraging students to think about their learning and to generate and act on ideas to improve their learning. Peer assessment is another way of providing students with quick and frequent feedback. This action also supports the development of students' social and co-operative skills. Studies of peer assessment indicate that students find the feedback of peers to be useful and reliable (Falchikov, 1995). Comments from students suggest that they perceive peer assessment as providing them with insights into how others approach, think about and complete a task and that the process of applying the criteria to others work helps them clarify the criteria and to reflect on their own work. However, students are sometimes reluctant to assign grades to their peers (Falchikov, 1995).

In summary, formative assessment is described in the literature as having three main aspects: eliciting the information, interpreting the information and taking action to improve the students' learning. Of these, the 'taking of action' is the most important aspect to distinguish formative assessment from summative assessment. In the following sections, different debates within the literature on formative assessment are discussed.

2.2 FEEDBACK AND FORMATIVE ASSESSMENT

As action is central to the definition of formative assessment, the need for feedback and dialogue between teachers and students is essential. Much of the research on feedback to-date has been based on stimulus-response learning theories (Torrance, 1993). In this case, the research has been on feedback providing information on whether ideas were right or wrong and the impact of the timing of the feedback

(Torrance, 1993; Sadler, 1989; Tunstall and Gipps, 1995). Within a constructivist framework, the notion of feedback is more complex as feedback has both informational and motivational effects (Falchikov, 1995). The process of formative assessment enables teachers to receive feedback on the effectiveness of the learning opportunities they are providing so they can modify them to optimise student learning. Teachers also provide feedback to students on what they know, understand and can do (Biggs, 1995; Radnor, 1994; Clarke, 1995). Raven (1992) argued that such feedback is important as people are often unable to perceive and identify their own unique qualities and abilities. Feedback can also identify areas of misunderstanding, through dialogue teachers and students can then generate opportunities for furthering student understanding. Such feedback enables students to direct their efforts more effectively (Sadler, 1989; Brown and Knight, 1994). When speaking of promoting multi-dimensional learning, Sadler (1989) quoted Ramaprasad (1983) in defining feedback as 'information about the gap between the actual level and the reference level of a system parameter which is used to alter the gap in some way' (Sadler, 1989). An important feature of Ramaprasad's definition is that the information is only considered as feedback when it is used to alter the gap.

This view of feedback implies students take an informed role in their learning. It implies teachers have a knowledge of the content to be taught, an understanding of how students are likely to learn it, a knowledge of the progression of ideas within the topic and are able to recognise where the students are in their development. It also implies they are able to use strategies to find out and develop students' ideas. This view of formative assessment is consistent with Vygotsky's notion of the 'zone of proximal development' (Vygotsky, 1978) and with Bruner's notion of 'scaffolding' (Bruner, 1986).

The centrality of feedback in formative assessment is also highlighted in the review by Black and Wiliam (1998). They discuss the need to address the effectiveness of feedback and refer to the work by Kluger and DeNisi (1996), who identified three levels of linked processes involved in the regulation of task performance. These three processes indicate the factors which impact, either negatively or positively, on the effectiveness of feedback. These three levels of processes were:

• meta-task process, involving the self. Feedback interventions that cue individuals to direct attention to self, not the task, tend to produce negative effects on performance.

> Teachers need to inculcate in their students the idea that success is due to internal, unstable, specific factors such as effort, rather than on stable general factors such as ability (internal) or whether one is positively regarded by the teacher (external). (Black and Wiliam, 1998, p. 51).

• task-motivation processes, involving the focal task. This type of feedback intervention draws attention to the task and is generally much more successful.

• task-learning processes, involving the details of the focal task. This type of feedback intervention draws attention to the details or characteristics of the task. Feedback appears to be less successful in heavily-cued situations and more effective in situations requiring higher order thinking.

Perrenoud (1998), in responding to Black and Wiliam (1998), takes a broader view on feedback by stating that:

all those evaluations are formative which contribute to the regulation of an on-going
learning process (p. 85).

What is required is regulation of learning (Perrenoud, 1998) and feedback is but
one source of regulation. Regulation differs from feedback in that the consequences
of the feedback is taken into account, not just the intent and practice of feedback. The
mere presence of feedback is not sufficient to improve learning. The effectiveness of
feedback is dependent on how the learner constructs and responds to the feedback. The
construction and response may not be as intended by the person giving the feedback.

Whilst it is simpler to observe the practice of assessment, it is more difficult to
discern its effects and consequences as is needed to research regulation of learning.
Perrenoud (1998) asserts that rather than being solely concerned with the formative
assessment practices of teachers, we need to conceptualise and observe more widely
the process of regulation at work in classroom situations. Such a regulation of
learning would be the regulation of cognitive processes, and the mediation and
interventions required to do so (Perrenoud, 1998). Perrenoud distinguishes two levels
of management of situations which favour the interactive regulation of learning
processes:

• the setting up of such situations through much larger mechanisms and
classroom management.

• the interactive regulation which takes place within didactic situations.

In summary, feedback and regulation are key aspects of formative assessment.

2.3 FORMATIVE ASSESSMENT AND ITS RELATIONSHIP WITH LEARNING AND TEACHING

Formative assessment is an integral part of teaching and learning. The current
definitions of formative assessment can support either behaviourist or constructivist
views of learning (and teaching) (Torrance, 1993; Sadler, 1989; Tunstall and Gipps,
1995). Behaviourist views of learning imply that knowledge can be divided into a
hierarchical set of discrete packages which teachers can teach sequentially and which
learners learn by mastering progressively more complex ideas. Mastery learning
practices are linked with this view. They assume that learning is improved if
students are aware of a teacher's goals and the outcomes which indicate they have
attained the desired knowledge. Within this view of learning, formative assessment is
a process of checking to see if students have achieved the lesson's goals and
providing feedback to students to help identify deficiencies.

Constructivist views of learning (Bell and Gilbert, 1996, p. 44) conceptualise the
learner as constructing their understandings from incoming stimuli and existing
knowledge. Therefore, learning can be seen as conceptual development (Osborne and
Freyberg, 1985, p. 82). A learner's prior knowledge is recognised as influencing the
development of their new understandings. This means that learning is not viewed as
proceeding along a single predetermined path to a pre-specified endpoint. Within this
framework, formative assessment is a process of examining the interaction between
the students' knowledge, skills and attitudes and the learning activity (Meltzer and
Reid, 1994). Further, when knowledge is viewed as socially constructed the roles of
the social context and social practices are important. Assessment processes are
viewed as social processes. In particular, the teacher's role involves providing
learning opportunities, introducing new ideas and interacting with students to
support and guide their learning. Learning becomes a process of appropriation and

negotiation of knowledge in a social context. Within this framework, the process of formative assessment is one of teachers and students recognising what a student currently understands and identifying what he or she could achieve next (Torrance, 1993; Meltzer and Reid, 1994; Clarke, 1995; Dassa, Vazquez-Abad and Ajar, 1993). This form of formative assessment is integral with teaching and learning (Meltzer and Reid, 1994; Harlen and James, 1996) and is more compatible with the notion of 'assessment while teaching', in which teachers routinely engage (Newman, Griffin and Cole, 1989). Teacher-student and student-student interaction are an essential part of the process, not separate from it (Torrance, 1993). Such assessment is jointly accomplished by the teacher and the student (Pryor and Torrance, 1996) and assessment information can be considered to contain information about the teacher and the students (Filer, 1993). Communication, feedback or dialogue, between teacher and student are essential in this case (Boud, 1995; Clarke, 1995; Filer, 1995; Sadler, 1989; Tunstall and Gipps, 1995; Torrance and Pryor, 1995). These communicative activities are not those found in traditional classrooms were learning is viewed as a transmissive process. Rather the inclusion of these activities suggests that to undertake formative assessment in the classroom, teachers must adopt a substantial change in their classroom pedagogy (Black and Wiliam, 1998).

A third perspective on learning is that of the sociocutural views of learning, which take into account the role of the social and cultural contexts in the learning process. These views are increasingly being used to theorise about learning in science (Driver, Asoko, Leach, Mortimer, and Scott, 1994) and form the underpinnings used to theorise the research findings documented in this book. A full discussion of sociocultrual views of learning and formative assessment are to be found in chapter 7, based on the findings reported in chapter 3-6.

2.4 THE ROLE OF STUDENTS AND TEACHERS

The conceptualisation of formative assessment discussed here implies different roles for teachers and students than traditionally found in classrooms. It implies a more active role for students. Clarke (1995) noted that what is assessed and how it is assessed is an important indicator and determinant of the these roles. In particular, if the teacher makes unilateral decisions about a student's learning, this sends particular messages to the students. When exploring the use of an assessment 'conversation' in the creative arts area, Radnor (1994) concluded that the position of the teacher as 'knower' needed to be temporality suspended if effective formative assessment was to take place.

The use of student self-assessment as a form of formative assessment also implies a shift in the traditional roles of teacher and student in the classroom. It encourages students to 'become insiders rather than consumers' of assessment (Sadler, 1989). It supports greater student autonomy and responsibility for their learning (Radnor, 1994; Willis, 1994). Fairbrother (1995) found that when students were involved in self assessment they were able, and more prepared, to comment on a teacher's assessment.

Two actions lie at the core of formative assessment, and both are student actions (Black and Wiliam, 1998). The first is the students' perception that there is a gap between the desired goal and her or his existing knowledge and skills. The second is the action taken by the student to close the gap (Ramaprasad, 1983; Sadler, 1989). Many factors can influence whether a student does or does not do these two activities, including whether students have a task or performance goal orientation; learners' beliefs about their own capacity as learners; the degree of risk taking involved for students; students' reflective habits of mind; and the degree of engagement with learning tasks (Black and Wiliam, 1998)

2.5 THE IMPACT OF FORMATIVE ASSESSMENT

Formative assessment, like other forms of assessment, has an impact on the teaching and learning in the classroom. Therefore, formative assessment (as one form of assessment) can be seen as influencing student learning in terms of the science knowledge and skills learnt. The notion of assessment impacting on learning will first be introduced with respect to summative assessment and then discussed with respect to formative assessment.

The impact of external summative assessment on teaching and on what and how students learn is well documented (Berlak, 1992a). Crooks (1988), in a survey of international literature, noted that classroom assessment had direct and indirect, intended and unintended consequences for learning. He found classroom assessment influenced what students considered was important to learn, their approach to learning and their motivation:

> Classroom evaluation affects students in many different ways. For instance, it guides their judgement of what is important to learn, affects their motivation and self-perceptions of competence, structures their approaches to and timing of personal study (eg. spaced practice), consolidates learning, and affects the development of enduring learning strategies and skills. It appears to be one of the most potent forces influencing education. (Crooks, 1988)

Others have found that assessment influences both what and how students learn (Boud, 1995; Harlen, 1995). For example, assessment can encourage students to take surface, deep or strategic approaches to learning (Crooks, 1988; Harlen and James, 1996). Deep learning approaches are considered desirable as they involve students in seeking to understand concepts and in making connections. Assessment, which emphasises understanding and the transfer of learning to new situations, has been found to promote deep learning (Crooks, 1988; Harlen and James, 1996). Assessment which emphasises the recall of isolated facts has been found to encourage surface learning, that is, learning in order to successfully complete the task.

Assessment can influence students' willingness and ability to engage in self-assessment (Boud, 1995; Crooks, 1988). Many writers argue that an important aim of education is to develop students' commitment to learning and their learning-to-learn skills. An essential aspect of these skills and attitudes is students' ability and willingness to engage in self monitoring (Sadler, 1989). Assessment can also influence students' willingness and ability to work and learn co-operatively (Crooks, 1988).

What science is being assessed impacts on the teaching and learning of science in the classroom. The science to be taught and assessed may documented in national curriculum documents or State 'standards'. In New Zealand, *The New Zealand*

Curriculum Framework (Ministry of Education, 1993a) calls for the curriculum and assessment to enable students to:

> develop (their) potential, to continue leaning throughout life, and to participate effectively and productively in New Zealand's democratic society and in a competitive world economy. (Ministry of Education, 1993a, p. 3).

This policy rhetorical statement suggests that education is seen as assisting students to develop the skills for life long learning and the knowledge, skills and attitudes to become critical and active participants in society. This curriculum document outlines areas of subject knowledge and essential skills and attitudes which relate to this aim. These include learning-to-learn, motivational and social attitudes and skills. The curriculum document *Science in New Zealand Curriculum* (Ministry of Education, 1993b) explicates these aims in relation to the learning of science. It details the science knowledge, skills and attitudes to be learnt, including those related to recognising and using science in everyday life. Because assessment influences what and how students learn, it is important that classroom assessment encompasses the range of all those learning outcomes which are valued in the curriculum, in a manner which supports those learning approaches considered most productive (Stiggins, 1991).

The educational goals described in the curriculum (Ministry of Education, 1993a) take time to develop and their sustained growth can be undermined by lack of consistent support from teaching and assessment practices (Crooks, 1988; Boud, 1995). Boud (1995) noted that assessment is part of students' everyday classroom experiences and so what they learn from one assessment is not interpreted in isolation. The message about what is considered important to learn and how to learn it, is interpreted in context, cues from the context provides students with clues:

> Students are not simply responding to the given subject - they carry with them the totality of their experiences of learning and being assessed and this certainly extends far beyond concurrent and immediately preceding subjects. (Boud, 1995, p. 37)

Like all assessment, formative assessment needs to communicate to students what is considered to be of value. It needs to encourage and support learning in a manner which consistent with how effective learning is viewed as proceeding. Claims for the value of assessment in improving learning are extensive but as Torrance (1993) noted much of what is claimed is based on rhetoric rather than on knowledge of what actually happens in classrooms. These claims derive from the assumption that accurate and representative information on student learning can be elicited if more effective assessment strategies are developed. They assume there is an automatic and simple link between assessment and the capacity to promote learning (Torrance, 1993; Savage and Desforges, 1995). Many of the studies which have successfully explored this link have been small scale, involving small groups or one-to-one interactions (Torrance, 1993). Studies into classroom life have suggested that the process is not simple (Bennet et al, 1984; Donaldson, 1978; Mehan, 1979). Torrance (1993) suggested that research into formative assessment should explore what is currently occurring within classrooms:

> It should enquire into what is currently happening and add to our capacity to act intelligently in difficult circumstances. This would be more ambitious in terms of what it sets out to accomplish, but more modest in what it might claim to achieve. (Torrance, 1993).

2.6 THE IMPORTANCE OF CONTEXT

The attention given to assessment to promote learning has led to a recognition of the importance of context in assessment (Berlak, 1992b). Research on assessment in general has found that students' performance is affected by the nature of the assessment task (Bachor and Anderson, 1994; Bachor, Anderson, Walsh and Muir, 1994; Crooks, 1988; Black, 1993). Formative assessment is also influenced by the classroom (Cowie, 2000). That is, the formative assessment process and outcomes are influenced by classroom factors such as how the students get on with the teacher, the layout of the furniture and the opportunities provided by the teaching and learning activities being used. Another classroom factor is that within a classroom the teacher holds most of the power as he or she controls what counts as valued knowledge and valued forms of interactions (Filer, 1993,1995; Radnor, 1994). Teacher questions play an important role in establishing and maintaining teacher dominance. The typical pattern of student-teacher interaction begins with the teacher asking a question. A student responds, the teacher evaluates the response and either probes further or moves on (Mehan, 1979; Edwards and Mercer, 1987). Other researchers have found that students often respond to the question they consider as implicit within a teachers' question rather than what is explicitly asked (Donaldson, 1978). Raven (1992) noted that peoples' actions are often determined by what they consider should be done and what it is appropriate for someone in their position to do. Students' perception of the teachers purpose for the task or interaction has been found to influence how they respond (Perrenoud, 1991).

Many researchers have commented on the significance of the social context in which assessment takes place. As the process of formative assessment often occurs within student/teacher interactions, it must be acknowledged that assessment has social functions and consequences (Hanson, 1993). Torrance and Pryor (1995) noted that the teachers within their study were sensitive to whether their feedback was private or public and to its impact on students' perception of self worth.

2.7 THE RELATIONSHIP BETWEEN FORMATIVE AND SUMMATIVE ASSESSMENT

Summative assessment has dominated research and development because of its status and the high stakes involved. As a result, the practice of formative assessment is not as well developed as that of summative (Black, 1993; 1995a; Black, 1995b). The current forms of external summative assessment do not provide good models for effective formative assessment (Black, 1995a). Although assessment strategies can serve formative or summative functions, it is the use of the information to improve learning which makes an assessment formative. Wiliam and Black (1995) claimed that all assessments have the potential to serve a summative function but some have the *additional* capability of serving formative functions (Wiliam and Black, 1995).

Some writers argue that an assessment can serve both the purposes of summative and formative assessment (Black, 1995a; Crooks, 1988). Given the need to ensure that teacher and student assessment loads remain manageable, it can be seen as desirable that assessment serve both purposes. It may be possible for teachers to review the evidence they have of a students' learning as a result of their practice of formative assessment in order to make summative judgements (Harlen and James, 1996; Wiliam and Black, 1995). Assessment practices, which have the potential for both summative and formative function, attend to the need for known learning goals, explicit criteria for judging success, feedback and for an opportunity to utilise the feedback. Summative assessment information, if used by a teacher to modify their teaching with another group of students, does have a formative function. If it is used by students to inform their learning approaches, it has a formative function for them. Black (1995a) noted that much of what is spoken of as formative assessment is in fact repeated or on-going summative assessment as no action is taken to inform learning.

Others claim that the formative and summative purposes of assessment are incompatible as they imply different roles for teachers and students (Gipps, 1994). Summative assessment requires the teacher to act as a judge of student learning and so it often involves a teacher stopping teaching to measure progress. Summative assessments tend to aggregate learning from disparate areas. Further aggregation occurs when the results of these assessments are recorded as marks or grades. In contrast, formative assessment tends to be continuous and informal, an integral part of teaching and learning (Cowie and Bell, 1996). In this case, the information about a student's understandings and skills does not need to be aggregated or recorded (Black, 1995b).

Summative and formative assessment also differ on the issues arising in current debates. Concerns, within summative assessment debates, are related to the consistency of the shared meanings of the assessment. As summative assessment results tend to be reported as grades, all students need to be treated in the same manner and the impact of the context minimised (Wiliam and Black, 1995). However, concerns within formative assessment debates are described as related to the consequences of the assessment for learning. This is reflected in a progression from a concern with technical issues to concern with the impact of the assessment on student learning (Sadler, 1989). Within summative assessment, the need for shared meanings has led to an emphasis on reliability and validity. Typically, reliability is defined as consistency among independent observations and validity as the extent to which an assessment measures what it sets out to measure. Within summative assessment, reliability is usually said to be necessary but not sufficient condition for validity because measurements may be reliable or consistent but still not be measuring what is of interest. Within formative assessment, the focus is on validity (Sadler, 1989; Harlen and James, 1996; Moss, 1994) and in particular on consequential validity. Consequential validity relates to the consequences of assessment on teaching and learning (Messick, 1989). Reliability is subsumed within validity in this case as it depends upon the self-correcting nature of consequent actions (Wiliam and Black, 1995). Essentially formative assessment interpretations and actions are always provisional, discussed and negotiated as part of the process of using the information.

A critical question in relation to the use of information for both summative and formative purposes is the confidentiality and potential harm of the information provided for formative purposes. If students are encouraged to take risks within the learning process and to be honest and open in their self assessment, there needs to be a clear and pre-arranged agreement about the possible summative uses of the information.

2.8 SUMMARY

Formative assessment involves the exchange of information between teachers and students about the students' learning. It is an essential component of effective teaching and learning. As a process, it is interactive and contextualised and it involves teachers and students eliciting, interpreting and acting on information about student learning. Ideally, it should support the development of students' personal, social and science development (Cowie, Boulter, Bell, 1996).

In the next chapter, one case study of formative assessment in science education is documented.

CHAPTER 3

A CASE STUDY OF FORMATIVE ASSESSMENT

The year was 1995 and one of New Zealand's active volcanoes, Mount Ruapehu, in the middle of the North island, was erupting. The timing could not have been better as Teacher 5 was starting a unit of work on 'Our Storehouse Earth' in which she planned for the Year 8 students to learn about the composition of the earth, tectonic plates, the cause of volcanoes, the composition of soil, rock types, and how rocks are formed. A case study of the formative assessment used by teacher 5 and her students during the teaching and learning of this earth sciences unit is documented in this chapter. It is one of the eight case studies in the research being reported in this book, with all eight case studies being documented in Bell and Cowie (1997, pp. 48-245). This case study was chosen to illustrate the data on formative assessment in the classroom which informed our modelling (chapters 5) and theorising (chapter 7) about formative assessment. Further illustrative data from the other case studies is given in chapter 6.

This case study is detailed, and therefore, long. It is felt that the detail is necessary to document, for the reader, all aspects of formative assessment: the actions of the teacher and students, the contexts in which it occurred, and the purposes for doing it. Hence, in this case study, both the social and cognitive aspects of formative assessment are documented.

This chapter is divided into the following parts:
- teacher 5 and her students
- the role of the researcher
- the setting
- the teacher's views of teaching, learning and assessment
- purposes for formative assessment
- the learning situations
- methods for eliciting formative assessment information
- interpreting the formative assessment information
- taking action
- summary
- three cameos of formative assessment
- summary and discussion

3.1 TEACHER 5 AND HER STUDENTS

In this case study, the classroom observations and interviews with Teacher 5 and seven of her students are described and discussed. Teacher 5 had had 17 years of teaching experience at the time of the research. Her teaching qualification was a four year BEd degree for primary (elementary) teaching. Her students were thirty Year 8 (aged 12-13 years) female students.

3.2 THE ROLE OF THE RESEARCHER

The researcher attended most of the lessons in this unit, with Teacher 5 and the students. The researcher's observations occurred over a period of six weeks with each lesson lasting one-and-a-half to two hours. The sixteen classroom observations for this case study took place between 12/10/95 and 24/11/95. An explanation of all the data codes is given in the appendix.

Participant observer

When in the classroom, the researcher acted as a participant observer, recording her interpretations of assessment in the form of field notes and noting documentary data, such as posters on the wall, students' workbooks and books used. The researcher did not audio-tape the lessons.

After the first lesson, the researcher spent most of the lessons with one group. On a few occasions, she was invited to join other groups, which she did for part of a lesson. On other occasions, when she was shown items or asked questions by other students, she went and worked with them in other areas. However, it was quickly accepted that she was part of one particular group. This group organised a place for her and commented to her that they enjoyed having her working with them.

During class discussion time, the researcher sat with her group and made field notes. The students sometimes looked at these and asked if she had recorded anything that they had said. She was included in looking at items which were passed around the class and in side conversations with the students beside her. During small group work, the students worked at their desks on the activities which had been assigned during the lesson. At this time, the researcher sat with her group. She sat at the desk of any student who was absent or she sat beside a student. She took an interest in what the students were doing, read reference books with them, looked for information for them, talked with them about what they were doing and enjoyed learning more about the topic herself. The students sometimes asked her questions. On one occasion, while she was discussing the requirements of a task with one student, another student interrupted the discussion and told them they were both wrong. The researcher made very few field notes at this time. When she did, she told the students what she was writing. She loosely monitored where the teacher was but as the teacher was often talking quietly to individuals, she was unable to field note the interactions between the teacher and the students at this time. During the classroom observations, the researcher was aware that she could only directly observe some of the teachers' formative assessment actions. The end of lesson discussions provided an opportunity to talk with the teacher about her assessment actions. The researcher's field notes are coded for example (T5/FN11/95b) to indicate these were field notes of the 11th lesson (FN11), taught by Teacher 5 (T5), in the second half of 1995, (95b).

End-of-lesson interviewer

The researcher was further informed by informal discussions with the teacher. Initially, the researcher talked informally with the teacher at the end of each lesson. These discussions often took place as or just after the students were leaving the room. They had the advantage that the lesson was still fresh in the teacher's and the researcher's minds. The structure of these discussions was informal, usually relating to specific episodes and specific children who were of note to the teacher and the researcher. The researcher's only planned question was to ask teacher 5 if she felt there had been any assessment in the lesson and if anything that occurred had surprised her. The discussions ranged in length from half to three-quarters of an hour, depending on the teacher's time constraints and the richness of the lesson. The researcher came to realise that these discussions were an important source of data on the teacher's interpretation of the assessment, which had occurred during the lesson. Teacher 5 described what she had assessed. She also described how she had done this, some of the judgements she had made about the students' learning and the actions she had taken. Teacher 5 later described these discussions as times when she was 'thinking aloud'. The researcher gained the teacher's permission to audio-tape the later discussions and it is these which form the basis for some of the data presented in this case study. There were 6 end-of-lesson interviews. The end-of-lesson discussions are coded, for example (T5/D11/95b) to indicate this was an end of lesson discussion after lesson 11, with Teacher 5 (T5), in the second half of 1995 (95b).

End-of-unit / end-of-year interviewer

The researcher interviewed teacher 5 at the end of the year, which was shortly after the unit of work had ended. During this interview, she briefly discussed the unit with the teacher. The interview is coded (T5/I/95b). She was also interviewed at the end of 1996 (T5/EOY/96). Seven students were also interviewed at the end of the unit (S55-57/I/95b) and these data are also reported in Cowie (2000).

3.3 THE SETTING

The first aspect of formative assessment that is important to note, is the setting. Teacher 5 taught all curriculum subjects (except technology) to the class in the same classroom. She was responsible for the learning, assessment and reporting programme used in the classroom. This programme included detailed written reporting on the students' science, personal and social development, a focus on student self-assessment and ongoing parent involvement in student learning and assessment (T5/I/95a).

The observed unit was just another 'Unit Study' - it was the researcher's presence that signalled the topic of study was science. Teacher 5 had taught the unit before and expected the students to enjoy it. She planned for the students to learn about the composition of the earth, tectonic plates, the cause of volcanoes, the composition of soil, rock types, how rocks are formed and the use of materials from the earth and had prepared worksheets to help with this. Activities for the unit were whole class discussions, six written and two practical tasks.

When the researcher arrived for the first lesson of the unit its the title, 'Our Store House Earth' was displayed on one wall along with posters and newspaper clippings about earthquakes, volcanoes and oil. This display was updated throughout the unit. Student interest was stimulated throughout the unit by the eruption of Mt. Ruapehu. Many students and the teacher visited the mountain during the unit and the mountain's ash cloud was often visible.

Resources for the unit were displayed on a bench in front of the teacher's desk. This positioning maximised her opportunities to observe the students as they worked with the resources (T5/D2/95b). Students brought books, photographs and artefacts (photographs, gemstones, necklaces and crystals such as amethyst) from home and added them to the resource table. The items brought in by the students provided the teacher with a robust source of information on what the student were interested in and the connections they were making (T5/D3/95b).

Classroom furniture and its arrangement both supported and constrained teacher assessment. Student desks were grouped and this allowed the teacher to observe the students at work. She considered this observation generated robust information because the students 'forgot' she was observing them. Five of the seven interviewed students indicated this was not the case. They claimed others worked harder and pretended to understand what they were doing while the teacher observed them (S53,53,55/I/95b). They did not like the teacher to look at their work when it was 'half done' because she might 'see something you don't want them to see' (S53/I/95b). Their concern may also have been because the teacher used observation as a summative assessment strategy and they were sensitive to what she might report to others.

Some students limited the teacher's incidental access to their written work by lifting their desk lids (T5/FN2,5,6,8/95b). One student assured the researcher this action was deliberate:

> I showed my friend and I quickly put up my desk when she came over so she wouldn't see it. (S53/I/95b)

Students covered their books and talked with their peers in a manner which restricted the teacher's access to their books (T5/FN4,5,7,8,10/95b). However, it seemed it was only the teacher's random access to their unfinished work the students disliked because they showed her their books and asked for her comments and help.

The desk arrangement facilitated peer and self-assessment through discussion, the sharing of resources, and the comparison of written work (T5/FN 5-14/95b). Students discussed ideas and then asked the teacher for help or looked at reference books together. They compared book work and then worked harder and changed or added to their work although the main focus of this assessment appeared to be the quantity - how many pages they had completed - rather than quality of work (S53,58/I/95b; T5/FN12,13/95b).

The teacher's planned assessment for the unit was a pre-unit questionnaire and four summative assessment tasks - a knowledge test, a presentation to the class of two of the questions they had explored, a student self-assessment and the marking of student books for content, and presentation. The students' presentations took place over the last three weeks of the unit.

A typical lesson

The observed lessons lasted one and a half to two hours. The lessons always began with a whole class activity followed by individual and small group work. During the whole class activity, the students sat on the floor, on a sofa or on chairs in an open space at the front of the classroom. The teacher sat on a low chair within the student group. The discussions lasted for three-quarters-of-an-hour to an hour. The teacher began the first lesson of the unit with discussion on 'Is the earth getting bigger?' based around photographs of ruins she had visited in Rome. Other lessons in the first half of the unit began with the teacher posing a question or, more usually, students talking about the artefacts they had brought from home. The discussions revolved around the layers in rocks and soil, the composition of soil, the colour and texture of rocks, the effect of light and water on the colour of rocks, crystals, gas, the nature of earthquakes, volcanoes and gemstones. The discussions constructed a weak boundary between the student's interests and experiences and school science. For the final three weeks, the students presented their answers to two questions they had explored to the class as part of their summative assessment.

The students worked on the set tasks and any questions they were interested in during the second half of the lessons. They moved freely around the room, working by themselves, talking in pairs or groups, looking at resources and sometimes going outside to complete a task. The teacher moved around the room. Sometimes she spent most of the session with one group, sometimes she circulated around the class and talked to most students.

The temporal context

A description of the setting also includes a description of the temporal context. The teacher stated she had a formed a 'picture' of the class as a group with well developed listening and questioning skills and that individual students had various levels of confidence, ability to express themselves and typical depth of understanding. She stated her perception that the students were able to discuss ideas, had influenced the nature of the learning tasks she had selected for the unit (T5/D4/95b), and as is evident later, the nature of her feedback:

> They know the expectation is that they will listen and that they are welcome to speak and ask questions. .... I suppose I know my class now. I have confidence that I and they have developed certain skills and patterns. .... On the whole I find this class at this point of time is good at listening. (T5/D14/95b)

The seven interviewed students said they expected the teacher to value certain behaviours and act in particular ways Their view was illustrated by the student who, when asked how she worked out what teacher considered important, said: 'I've sort of got used to what she thinks is important and stuff' (S58/I/95b). Knowledge of the teacher's usual actions was used to interpret her interactions and written feedback.

The interviewed students also indicated they considered some of their peers to be 'bright' and as likely to understand ideas, and others as able to be 'trusted' not to make fun of them when they asked a question. These perceptions were reported to influence their actions (S54/I/95b).

3.4 THE TEACHER'S VIEWS OF TEACHING, LEARNING AND ASSESSMENT

A further aspect of formative assessment is that of the teacher's views of teaching, learning and assessment. Teacher 5 described learning as an activity that involved individual students 'building' on their ideas and as the 'growth' of collective or group knowledge. Assessment was described as something done by teachers and students but 'on most occasions ... it's a combination of pupil and teacher identifying these things' (T5/I/95a).

The teacher's description of her role as a questioner exemplified her attitude to teacher assessment, she said:

> ... as a questioner I can generally find out anything I want to find out. ... If you take an interest in what they're doing, they are only too happy to explain and to share. They want to. [The children] enjoy that process and it is good for (them). Children with an idea are given the opportunity to talk about it. If they thought they had a problem, often the solution comes to them as they talk about it out loud.... the moment they start talking about it 'Oh I do know. I can do such and such can't I?' ... (T5/I/95a)

She considered assessment as 'something educators need to do to help with the next stage in the children's learning and meeting their needs' (T5/I/95a). For this reason, students were assessed at the beginning of a unit to find out what they knew, so she that could 'use what they know and build on that', they were assessed during a unit to find out if the teaching programme was promoting the learning she had planned for and so she could follow-up student ideas. She noted there was 'space within my unit to actually shoot off if anybody comes up with an idea' (T5/I/95a).

The teacher stressed assessment was a mutual responsibly. It was her responsibility to provide students with a range of opportunities for displaying what they knew and could do, and the students' responsibility to tell her if she was not meeting their needs (T5/D14/95b).

Assessment was also described as a process students needed to engage in for, 'identifying areas that they need to work on, ... identifying areas they want to work on and accepting areas they need to work on (T5/I/95a). It was one of her long term goals for student to learn to assess themselves as she considered this would enhance 'their own personal quality of life' (T5/I/95a). She commented that student self-assessment required her to 'shift' some of her power to the students so they could 'build up a responsibility' (T5/I/95a).

To summarise, the teacher's comments suggested she saw teaching as using what students 'already know' and 'building on that' (T5/I/95a) and assessment as integral to teaching. Her description of assessment as a teacher-student responsibility indicated she considered that teachers have limited access to student thinking. The importance she placed on student self-assessment (for students now and in the future) suggested she viewed students as active meaning makers. Her comments that students needed to identify what they wanted to work on suggested she viewed motivation as an integral to learning.

3.5 PURPOSES FOR FORMATIVE ASSESSMENT

In documenting the formative assessment that occurred, the purposes for doing it, need to be noted. Teacher 5 assessed her students with respect to their personal, social and science development (Cowie, Boulter, Bell, 1996). In general, her students' personal and social development were long term goals, while her students' science development goals were more likely to be short term ones associated with the unit or a lesson. Hence, this teacher had short and long term learning goals for the students, and therefore, long and short term purposes for the formative assessment she did in the classroom. The formative assessment of the personal, social and science development is detailed in the following sections.

Personal development

Within the classroom, personal development was conceptualised as those learning outcomes which relate to the learner as an individual, for example, their learning-to-learn, time management and self-assessment skills. Teacher 5 sought to promote these learning outcomes and formatively assessed her students' development of them.

Teacher 5 formatively assessed her students' time management skills. Near the end of the unit she systematically looked at the students' books. She said she noted which tasks they had completed and the quantity and quality of their work (T5/D14/95b), thereby assessing the students' time management of their learning activities.

Teacher 5 also formatively assessed her students' learning-to learn-skills. Teacher 5 said she considered that the ability to gather information from a number of resources was important. For this unit, 'Our Storehouse Earth', the teacher and the students had collected a large number of books to act as a resource to help the students answer the questions which arose within the unit. The teacher commented on the students' ability to gather information from books. She said:

> Some of these kids are really good at browsing through books. When you ask a question, ... as questions have come up there have been a number of children who go straight over there and they say, 'I've seen that'. J is really good. 'Look at this, that's

the San Andreas Fault'. She found the picture of that the other day because she had
seen it before. It had gone in and it had stayed there when she had been browsing.
(T5/D11/95b)

When she showed the class a video on volcanoes, she commented that she
intended this to be an opportunity for the students to gather information from another
source. She considered videos as an important source of information because of the
time students spend watching television:

> ... data gathering from another source. And there are lots of ways they can get
> information. Video is an important one, ... they spend a great deal of time in front of a
> television screen. (T5/D11/95b)

Teacher 5 intended for all the students to use a resource to research a question and
share their answer with the class. She explained that she considered explaining to
others was an effective technique for developing understanding:

> So one of the things is, it's an organisational thing coming through at the moment, is
> getting everybody to go and do some research and come back and share. When you
> actually have to look something up and you have to put it in your own words when you
> haven't got your notes in front of you ... that is often a way of internalising information.
> The understanding begins to develop. It really does develop or else you make a break
> through or something like that. (T5/D10/95b)

The researcher field noted that during this lesson that the teacher asked the class
who had researched the question 'Why is the top layer of soil darker?' Only six
students put up their hands. She commented on the need for them to do their own
investigations and not rely on others (T5/D10/95b). She therefore publicly and
formatively assessed this aspect of her students personal development.

Teacher 5 also provided opportunities for her students to develop their self-
assessment skills. She included a self-assessment in her end of unit assessment.
During one lesson, she formatively assessed the students' ability to assess their own
contribution to the learning and development of the class. She asked those students,
who considered they hadn't made a significant contribution to the whole class
discussions, to try to link some of the ideas the class had been exploring. She
commented on the student who volunteered to do so:

> I was pleased she acknowledged she had not said very much and was prepared to do
> something. (T5/D14/95b)

In this example, teacher 5 linked the skill of gathering data, a personal skill, with
the ability to share and discuss ideas, a social skill. These aspects of students'
personal, social and science development are conceptualised as interlinked, with many
opportunities for developing a student's personal development occurring in a social
context. The two aspects are interconnected and interdependent. In the instance above,
teacher 5 linked the development of her students' personal and social skills. She
commented on the variation in her students' research skills and their different
willingness to share their ideas:

> One of the things that I think has come through is that we've got some very able
> students who follow very well. They go home and they actually look things up. They
> can't wait to come to school. They enjoy sharing what they've found. There are others
> who go away and they look it up, but they come to school and they sit back and they
> wait. Then there are others who think, 'No, so and so will do it and I'll just listen when
> she tells us'. (T5/D10/95b)

On another occasion, she named students whom she considered possessed these attributes. She stated she assessed her student's ability to gather data from different sources and she assessed her students' ability to speak in the whole class situation.

In another example, teacher 5 stated that she assessed her students' ability to understand ideas:

> When we are looking at the use of the video, the assessment is on two levels. The information that they are getting, the understanding that's being developed and also their skills. This is another skill, gaining the information. (T5/D11/95b)

Here, teacher 5's assessment of her students personal development was linked with her assessment of their social and science development.

Social development

A second aspect of the teacher's purposes for doing formative assessment was that of promoting social development. Within this research, social development was conceptualised as the students' development of their skills of interacting with and working with others. The social development of her students was a long term learning goal for teacher 5 and hence provided a purpose for doing formative assessment. At this stage of the year, the teacher and the students had well defined expectations of each other. For example, over the period of the observations, teacher 5 only reminded the students three or four times to listen to each other. It appeared that the students very rarely failed to meet the teacher's expectations. At these times, she commented on the importance of listening to learn and of being courteous. It is interpreted from this and her previous comments that the teacher valued and formatively assessed her student's listening skills. At this stage in the year, she considered this skill was well developed and so it was not a main focus of her assessment (T5/D14/95b)

She continued, explaining why she considered that listening was an important skill:

> I actually think that listening to people, following conversations, is something we get better at. I think this is a way that we can help children, by giving them the opportunity to practise these skills. (T5/D14/95b)

And on another occasion:

> ... I suppose I know my class now. I have confidence that I, and they, have developed certain skills and patterns. .... On the whole I find this class, at this point at time, is good at listening. ... The organisation stuff is critical. ... (T5/D10/95b)

Science development

Another purpose for teacher 5 doing formative assessment was to assess her students' science development or science learning. Students' science development is conceptualised as including the students' science content, science processes and science context development (Cowie, Boulter and Bell, 1996). Within the unit described in this case study, the teacher emphasised science content and contexts over processes because of the topic.

When she spoke of the learning activities she provided (for example, whole class discussion, a video, task sheets, handouts and investigations) to mediate the learning of science within her classroom, she identified two aspects which she assessed. These were whether the students were developing an understanding of the ideas, that is, she assessed the science content, and whether they understood the task. For example, during small group work she assessed her students' ability to distinguish between the continental crust and the tectonic plates and their ability to complete the task of colouring the plates. (T5/D11/95b)

Teacher 5 was concerned with formatively assessing the learning and progress of the class as a whole, as well as the learning of individual students within it. She spoke often of assessing the whole class for the level of knowledge and interest within the class. She monitored this in order to time her input of new ideas. During many of the informal discussions, she spoke of this formative assessment of the learning of whole class:

> ... after the video, when we went through the questions. ... what I was doing with that, was trying to get a feeling about where we were. At the beginning of the unit, people would ask questions and we just didn't have answers to them. Now, those questions are still there and lots of people are putting up their hands. That was a general indication. (T5/D11/95b)

In the end of unit interview, she stated again that one of her main purposes for assessment was to monitor and promote the growth of knowledge within her class (T5/I/95b).

Teacher 5 also talked of assessing her students' ability to link what they were learning with their everyday lives, that is their science context development. She expressed her pleasure when she assessed that a number of students were linking their everyday experiences with the science in the classroom:

> The grouped work at the beginning ....We found they've got a lot of everyday experiences which I don't think they would have related to science before this. I thought that was quite valuable in that they talked about everyday things while here we were talking in a science lesson. I think quite a few of them .... might have come a bit closer to realising the relevance of what we were talking about to their everyday life. (T5/D11/95b)

Divergent and convergent assessment

Another purpose for formative assessment for teacher 5 was related to her planning for her students to come to know and understand certain ideas, as well as planning for them to pursue ideas which interested them. Her assessment of whether the students had come to know and understand the ideas she intended constituted her convergent assessment (Torrance and Pryor, 1995). Her assessment of what the students had learnt by following up the questions they or other students posed, constituted her divergent assessment.

Teacher 5 stated that she planned for the students to learn some specific concepts and she planned to create the space and opportunity for her students to pursue questions of their own. She indicated this during an end-of-lesson discussion and during the end-of-unit interview. In the discussion she said:

> ... the children are interested, they are attentive, many of them are asking very good questions and making excellent observations. They are following things that are occurring to them which have come up from the study, things which haven't come directly from me. (T5/D11/95b)

During the whole class discussions, the students in this class posed many questions of their own. Teacher 5 identified some of these as 'good' questions, focusing the students on them and recording them in her work book.   Teacher 5 revisited these questions during a lesson towards the end of the unit (T5/FN11/95b). For example, one 'good' question was 'Why are there more volcanoes in the pictures of dinosaurs? Were there more volcanoes then?'. Teacher 5 and a group of seven students investigated this question during one lesson (T5/FN3/95b) and two students pursued this question during class time over a period of two weeks with occasional help from the teacher. Teacher 5 stated that two other students had written up their answer to this question in their books (T5/I/95b).

Another example of a teaching activity that enabled teacher 5 to make divergent formative assessments was towards the end of the unit, when the teacher asked each student to present three tasks of their choosing to the class. The teacher intended that this provide an opportunity for the students to demonstrate to her and their peers what they had learnt. Most students presented one of the set tasks, usually the volcano activity. Two students presented more detailed information on the tectonic plates.

Another example of divergent assessment was the teacher's focus on the students making connections and linking ideas to everyday contexts because her purpose was to determine what sense the students were making of the learning tasks. As the teacher planned for the students to do this, it was also a feature of her convergent assessment. These two approaches to assessment were interlinked.

Teacher 5's divergent formative assessment tended to be of the science ideas her students had developed as a result of pursuing questions which were posed by them or other students in the class, and of how they linked their scientific ideas to their everyday lives. Her convergent formative assessment tended to be of the students' personal and social development. For example, she intended the students to develop further their skills of locating information and she formatively assessed to see if they were using these skills. Convergent formative assessment also included many of her short term goals (for the unit or for the task) for the students' science learning, as she assessed their engagement with a particular task and their development of an understanding of a particular scientific concept.

Teacher 5 indicated that she thought that there was an impact on the learning of the divergent and convergent formative assessment. For example, she considered that specifying her summative assessment requirements too soon could affect the quality and depth of her students' learning. Within the unit which was observed, she had planned for her students to explore questions which interested them. She stated that seeing her students happy to be doing science, asking questions and suggesting answers was one of the joys of teaching. She considered that when she told the students of her summative assessment requirements, they shifted their focus slightly. They then wanted to know how, what and when she wanted them to learn:

> What I did notice though ... there was, to me, a slight change in their attitude when I issued the sheets about assessment. They have been really enjoying going through this unit. The moment assessment is mentioned, there is a shift in their attitude, the way they feel. All of a sudden they want to be specific. When is it due? What is due? They need to know. ... it will be interesting to see if that happy, happy attitude, happy to be doing science attitude, just shifts a gear, now that they know what the assessment is. This is why ... I do always like to tell them the purpose of the study at the beginning of a unit but I don't always like to tell them how it's going to be assessed. Not until they need to know because I want them to get into it and start enjoying and gaining. I want to see them growing with no, what they consider, ulterior motives. (T5/D11/95b)

Teacher 5 wanted to encourage divergent learning. She stated that she felt that introducing her convergent assessment tasks too soon, albeit convergent summative assessment, encouraged the students to become more convergent in their learning. Her summative assessment consisted of requiring the students to complete the set tasks and investigations, their presenting three items of interest to them to the class, completing a self-assessment and a test on scientific content. The researcher observed that the students in the group, of which she was part, became more focused on completing the tasks after they learned of the teacher's requirements. They discussed the number of tasks they had done and compared their work. However, it was difficult to determine whether this was because they restricted the scope of their interest and learning or because they had a time deadline to meet.

In summary, Teacher 5's purposes for learning and assessment related to the students' personal, social and science development. Teacher 5's purposes for her students' personal and social development tended to be long term, her goals for the year. Her purposes for her students' science learning were usually associated with the unit or lesson, although she intended that the students link their school science with their everyday experiences within of all the science units. Teacher 5 used both convergent and divergent formative assessments.

3.6 THE LEARNING SITUATIONS

The fourth aspect of documenting formative assessment is that of the learning situations in which formative assessment was done. The students in the class observed in this case study were involved in three main learning situations. These were the whole class discussions, small group work on tasks and questions, and watching a video. Each of these learning situations produced a particular formative assessment environment. Within each of these situations, teacher 5 gathered information from the class and from individual students. She gathered the assessment information using questioning, listening and observation, each of which provided her with different forms of information. In these situations, she planned for and then elicited information, she created opportunities which facilitated her gathering of formative assessment information and she took advantage of opportunities as they arose. In each of the three learning situations, how, what and from whom she gathered information varied. In the following analysis, each learning situation will be analysed using these features.

Whole class discussions

In the unit of work described in this case study, teacher 5 started each lesson with a whole class sharing and discussion time. The whole class discussions provided the teacher with the opportunity for informal, on-going formative assessment of the class and individual students.

In the whole class situation, she was able to observe who contributed and listened to the discussion and the range and depth of the ideas. She was also able to question the students in order to probe their understanding. As the topics of the discussions were determined by the what the students brought to class, it is interpreted that this situation provided her with opportunities to collect information on the students' interests and understanding. The students also questioned each other and this provided the teacher with further insights into their thinking. As the students' existing knowledge and the links they were developing were articulated, the teacher was able to assess their understanding and learning. She was able to undertake divergent assessment. For example, when the class discussed why the top layer of the soil is usually darker, one student suggested that it was sunburnt (T5/FN10/95b). This student returned to this explanation in subsequent discussions on soil layers. When teacher 5 introduced a topic for discussion, for example the composition of soil (T5/FN10/95b), she was also able to undertake convergent assessment of individual's science explanations, their confidence and ability in speaking within a group.

The whole class situation also enabled her to formatively assess the scientific understandings which were developing within the class. For example, on one occasion, she put up on the board, a summary of the questions the class had generated during previous discussions. She asked the students to indicate if they considered they could now answer these questions. She stated she often used this technique to assess the general level of understanding in the class:

> ...again I just asked some general questions. ... when you ask general questions you can usually gauge, by the number of children who respond, how well the information has gone in. I think that I use that quite a lot. (T5/D11/95b)

In this manner, she obtained a 'general impression' of the students' or class's knowledge at that point.

Teacher 5's opportunity for eliciting information on all the members of a class of thirty students, using whole class discussion, was constrained by many factors. One of these factors, which was highlighted by this teacher on several occasions, was that it was only possible to elicit information from students who contributed to the discussion. Teacher 5 stated she considered that only highly motivated and confident students spoke in this situation:

> ... not everybody will get assessed in the general stage. It is only those highly motivated kids who are good at talking, the confident ones. (T5/D11/95b)

By using whole class discussion to elicit information on her students' learning and interests, the teacher appreciated that she only collected information on a random sample of her students - those students who were confident and highly motivated. She stated that if she felt it was essential for every child to speak, she used small groups:

> ... If I wanted every one to have a say I would use small groups. In a whole group, there is a danger the dominant kids, the knowledgeable kids, the confident kids, will do all the talking. If they're giving good information, which is sensible, makes sense and leads to further discussion then my job is to encourage as many different kids as possible to take part in the discussion. There are some children who resist it absolutely. L does. The resistance from her is amazing. (T5/D14/95b)

Teacher 5 described a strategy she used to complement the random gathering of information by checking whom she had information on, and then systematically eliciting formative assessment information on the other students :

> So what I've got to do at this stage, is sit down ... and go through the roll and say: Right, I'm happy about these people here. I know that they are good at talking, they've got the words, the vocab, they've got sound, sensible ideas. I haven't heard anything from these children, this group of kids so I've got to get to them. And then there are other children I'm not certain about and I'll have to go to them. That's what I intend to do. ... you've got to go to the groups, to validate it (the information) in a way. ... you go to the groups to follow up on the kids who didn't have any input in the general discussion and find out where they are. (T5/D14/95b)

For teacher 5, the issue of 'validation' of the information gained through whole class discussions involved another issue. When students were asked to volunteer to answer questions in the whole class situation, it was not possible to determine whether those who didn't answer did so because they didn't know the answer or because they lacked the confidence or the desire to respond.

Teacher 5 used a formal written summative assessment at the end of the unit in order to elicit information on all students:

> These other less confident kids, the ones who are definitely gaining what I want them to gain, I'll pick up on them when I formalise my assessment. This is that each child will present three tasks ... they will do a written assessment .... (T5/D12/95b)

Using a global, informal and random technique, such as formatively assessing students' contributions during whole class discussions, was a technique which was useful for gathering specific information about some students and a general impression of student knowledge. Teacher 5 also elicited more detailed information using more targeted techniques. She often discussed her technique of moving from gaining a general impression to more detailed and specific information about individual's understanding with the researcher (T5/FN6, 11,12/95b).

In summary, teacher 5 was aware of the limitations of using whole class discussion as a technique for gathering detailed and specific information on all the students. However, it was a technique which enabled her to collect divergent and convergent assessment information on student interests and ideas in the form of explanations. She was also able to assess student confidence and speaking ability.

Watching a video

The second learning situation used by teacher 5 was watching a video. During this unit, the students watched a video on volcanoes and the teacher spoke of the formative assessment information she had elicited in this situation. For example, while the students were watching a video she asked three students to move forward. She described this episode during the end of lesson interviews:

> ... And there were three children who were really struggling. There were too many distractions between them and the video. That's why I moved them forward. (Right, so how did you pick that up, that they were struggling?) Because I watched them. (T5/D11/95b)

In this instance, she was observing the class and she identified individual students who were having difficulty with the task. She moved these students, an action which she considered would reduce the distractions for them.

Small group situations

The third learning situation, which teacher 5 used to elicit formative assessment information, was small group situations. The class spent a part of most lessons working in small groups on the assigned tasks and worksheets. During the small group work, teacher 5 circulated around the class, talking to the students and looking at their books. The atmosphere at this time was relaxed, with students working on the assigned tasks or on questions which interested them. This created an environment in which the students were responsible for their learning and were observed to approach the teacher and show her their work or ask her questions. This situation created many opportunities for informal, on-going formative assessment by the teacher and the students. At this time, teacher 5 was able to undertake convergent and divergent formative assessment of the students' learning by observing their work, questioning them and listening to their answers and their questions. She was able to deliberately assess some aspects of the student learning and to notice others. For example, during one lesson she systematically assessed how close each student was to completing the set tasks by looking at their books (T5/FN12/95b). During another lesson, she handed out a task which involved the students identifying and naming the tectonic plates. She then assessed the students' progress with and understanding of the concepts in this task (T5/D10/95b). After another lesson, she told the researcher that some students didn't understand the concepts associated with soil, land forms and ore (T5/FN10/95b). The teacher assessed the students' understanding of concepts and the requirements of the task by talking to them and looking at their books.

Teacher 5 observed the students during the small group work. For example, during one lesson, the researcher worked on the floor with two students who were endeavouring to classify rocks. The students had noticed a link between some types of sedimentary and metamorphic rocks. When teacher 5 came over to talk to them, they explained their idea to the teacher. During the end of lesson discussion, teacher 5 asked the researcher if she had contributed this particular idea as presumably, she had observed the interaction between the researcher and the students (T5/FN7/95b). On other occasions, teacher 5 was observed by the researcher to identify students who were looking unhappy and then to spend time with them.

Teacher 5 appeared to collect a considerable amount of formative assessment information through informal observation and discussion with students during small group work time. Some of it she planned to collect. Other information she collected as the opportunity arose. She collected convergent and divergent assessment information in the form of student questions and verbal and written explanations.

In summary, the learning situations and activities which teacher 5 provided appeared to structure the type and scope of the formative assessment information she was able to gather. Each of the three learning situations, used by teacher 5 within the observed unit, enabled her to collect different formative assessment information using questioning, listening and observation.

3.7 METHODS FOR ELICITING FORMATIVE ASSESSMENT INFORMATION

A fifth aspect of formative assessment worth noting is the methods used by the teacher to elicit information. Teacher 5 used the methods of gathering assessment information which were described by the ten teachers in the first phase of the research (Cowie and Bell, 1995). In particular she used observations, questions and listening to students.

Observing students

Teacher 5 used observations to collect formative assessment information in all three learning situations. Observation appeared particularly useful for informally assessing the social development of a class; how the class was working; and whom was working with whom. The data in this case study also suggested it was useful for formatively assessing who was contributing what ideas and for gaining a 'general impression' of the level of scientific knowledge within the class. Looking at student books, both systematically and in response to student requests, teacher 5 obtained more detailed information on individual student thinking as the students were recording their own ideas in their books. Through observation, teacher 5 was able to deliberately collect particular formative assessment information and to gather other information by creating or noticing formative assessment opportunities.

Teacher 5 consciously organised the physical environment to facilitate her opportunity to observe the students during their learning. She positioned the resources, which the students used during the unit, along a bench in front of her desk. The students gathered at this bench to look at the resources and usually talked to each other about their ideas while they were doing so. By positioning the resources in front of her desk, the teacher maximised her opportunity for observing these interactions. On one occasion, at the end of a lesson, she spoke with obvious pleasure of observing one student helping another. She described this a something she strove to promote but didn't often have the opportunity to observe:

K and A wandered to the books. K had asked the question 'What is an oil rig' She had no idea what an oil rig was. A was picking up some books and saying 'There are some books here about oil rigs K'. ..... It was really good that A helped her. ... it's not a surprise that these kids do it, I mean, that's what I hope for all the time. I say to them, help one another, ask one another questions. If you've got something that will help somebody else, give it to them, tell them, let them have it. I say these things, but then, sometimes I feel I'm waiting in vain to see it happen. ... I suspect that it does happen a lot, more than I think it does. You've just got to be in the right place at the right time to see it. ... And having the resources right in front of the teacher's desk helps ... (T5/D14/95b)

Teacher 5 was sensitive to and noticed what was happening within the classroom, and was able to 'be in the right place at the right time' to gain insight into student learning as it was developing. For example, she was fortunate to be watching one student while she was replying to a question:

She volunteered some information to try and relate the stuff up there to 'The Storehouse Earth' theme. She was on the right track. She felt good about that ... I was standing there beside her and I could see her face and reactions. (T5/D14/95b)

During this episode, she was also able to observe another student who indicated by her body language that she also understood the ideas involved:

There are still some kids like A ... (she lacks) the confidence to speak up ... in a class situation, I could see from her reaction ... that she was saying quietly, while J was speaking, ... she was on the right track as well. She was feeling a bit more confident but not confident enough. (T5/D14/95b)

Teacher 5's observation of student learning was often supplement by or occurred concurrently with questioning and listening to students.

Listening to and questioning students.

Two other methods used by teacher 5 to elicit formative assessment information were listening to and questioning students. These techniques were usually used in conjunction with each other, although teacher 5 also spent time listening to the student-to-student discussion during the whole class discussions. Teacher 5 was observed to use listening and questioning to elicit formative assessment information on both the class and on individual students. These methods provided general and detailed information on the students' learning, what they know, understand and can do, depending on the situation and the questions asked. These methods were used for the convergent or divergent formative assessment of student learning. Their use was both planned for or arose as a consequence of the structure of the learning situation or the organisation of the environment. Either the teacher or the student was able to initiate the interaction.

Formative assessment information was gathered by teacher 5 by listening to, questioning and observing students. How the assessment information was gathered was determined to some extent on what and why teacher 5 wanted the information. Consideration needs to be given to how the quality of the assessment information was affected by how it was gathered.

Influences on the quality of assessment information

Teacher 5 considered that many factors influenced the type and quality of the formative assessment information she was able to gather from her students. When the method of gathering the information involved dialogue between her and the students, teacher 5 stated that she considered a students' motivation and confidence, their mastery of the language of the subject and their ability to articulate their ideas could influence the quality of the information she gathered. The first factor, that of student confidence in speaking, was discussed earlier. The second factor was a student's ability to articulate their ideas. For example, students were required to present to the class a task which they had completed. One student volunteered to do this:

> ... So I think A   was still on the surface of her piece of work. Maybe she had been relying on having K   being there to help her out and she had been encouraged to do it by herself. Maybe A is just not very good at explaining her ideas sometimes.   ... I need to look further at her.   ... to see what (she can do) and to take notice of the fact that when those others tried to help her out, she said, 'Oh yes' and she tried to chime in over the top of them. ... but if she was just left to her own devices she wouldn't do herself justice. (T5/D12/95b)

On another occasion she said of this student:

> There are still some kids, like A   ..., (who lack) the confidence to speak up, ... because A   knows that she struggles with a lot of the technical stuff and she doesn't want to use wrong words ... (T5/D14/95b)

Teacher 5 indicated that she was concerned that the student might not do herself justice in these situations. It appeared that the teacher attributed this to the students' confidence and ability to use vocabulary, and to articulate her ideas. She stated she needed to be aware of these issues and to provide the student with other opportunities to demonstrate their learning. She spoke of this again in terms of the variation in confidence and learning styles of her students:

> ... So you see it comes back to learning styles doesn't it? ... some kids prefer to be in small groups and we've been in a large group a lot of the time. You see that (being in a large group)would have scared a lot of my Form 2 kids off. ... Some kids prefer to be in small groups ..... it is very important to provide a lot of different sort of situations. (T5/D14/95b)

Student confidence was also associated with their opportunity to be assessed. Teacher 5 commented on the necessity she felt to monitor the influence of a student's confidence on the student's opportunities to be assessed and the impression the student made on her as the teacher:

> ... The confidence thing is important because often children do know the answer but they just do not have confidence to say it. It is very easy to over-rate children's knowledge because they are very confident. I have one student who is very confident but her knowledge is still very limited. On first impressions you might think 'Oh yes, she's good, she has a lot to say and she's very interested'. But she wastes time when speaking and doesn't say anything specific ... (T5/D14/95b)

Teacher 5 was aware that when she and a student were talking with each other, it was necessary that both of them to construct similar meanings for questions and answers. On one occasion, teacher 5 discussed an incident during which she had asked a student a question and the student had not been able to respond. She said:

> When I asked her a question, she didn't answer it. I later realised that she did know the answer. Maybe she didn't realise what I was trying to get at. It wasn't until I said 'Was the bottle this shape?'(she motioned with her hands) ... I've got to make sure that what I'm saying is what they hear and what I mean is what they understand. Often we assume they understand and they don't. .... I have done a unit on the responsibility of the speaker and the responsibility of the listener. ... They (the students) don't consider that the listener ... has got to concentrate, focus and imagine what the implications are for them and then ask a question. (T5/D10/95b)

In this instance, although teacher 5 acknowledged her own responsibilities for ensuring that the student understood what she required, she stated that both the speaker and the listener had responsibilities. During a previous interview, she had elaborated on what she considered her students' responsibilities for assessment:

> They all have the opportunity of adding questions on there (on the board). ... they can get somebody else to ask their question too. ... At other times of the day, I talk to the kids about their responsibility. I tell them, 'I'm not a mind reader'. If they've got a question, which I haven't answered ... and maybe I should have answered or I should have had the question answered in some other way, if I haven't done that, they've got a responsibility in the process, they've got to help me with this. (T5/D14/95b)

It is interpreted that she considered that any understandings were mutually constructed and that both the teacher and the student(s) had a responsibly for ensuring this was accomplished. Interpreting a question was problematic for this teacher as well. During one lesson, the teacher and the researcher formed two different interpretations of a question which a student asked (T5/D10/95b).

The influence on the assessment information of the techniques used to gather it, suggested that formative assessment information is, in part, a product of the techniques used to gather it. Teacher 5 suggested that what information she gathered, particularly that which was gathered through her questioning and listening to a student, was a product of the interaction between her and the student. This suggests that the interaction or method of gathering the information would impact on the honesty, accuracy, representatives and usefulness of the information to both the teacher and the student(s). In the best case scenario, the teacher and the student(s) would co-operate to produce the formative assessment information. This would help ensure that the formative assessment information is accurate and useful to both, something which is necessary if either are to effectively act on the information. In the class observed during this case study, the students often volunteered information. For example, teacher 5 spoke of finding out about student learning because the students came and told her:

> They are following things that are occurring to them that have come up from the study, they haven't come directly from me. ...(So when you say they are following, how do you know they are following?) Because they ask a question, they come back a little bit later and say: 'Look what I've found about what I was asking you about'. (T5/D11/95b)

The following-up could be as a consequence of the climate in classroom, the level of interest in the topic or the learning activities which the teacher used. There were many opportunities for the teacher and the student to interact informally. The teacher also actively encouraged students to take responsibility for their learning and to contribute ideas. Teacher 5 thought that some students enjoyed sharing their ideas (T5/D11/95b) and she considered that she gained insights into the students' learning through this process:

> (A) she had some input today. She has been quiet until today. I think this is a sign. When you start to make connections you want to verify them and so you ask for this. I think this was what she was doing. (T5/D12/95b)

The student was prepared to share her developing understanding. Hence, teacher was able to gather a relatively accurate insight into what the student was learning and how the student perceived her learning as progressing.

In this section, teacher 5's techniques for gathering different assessment information have been linked to her purposes for gathering it. The possible limitations and impact of some different techniques have been discussed. The interpretation of assessment information will be discussed in the next section.

3.8 INTERPRETING THE FORMATIVE ASSESSMENT INFORMATION

Another component in the formative assessment process to be noted was the interpretation of any assessment information before action is taken on it. The literature suggests that assessment is usually norm-referenced, criterion-referenced or ipsative assessment (Wiliam, 1992) . Teacher 5 appeared to make ipsative and criterion-referenced interpretations of the assessment information she collected during the lessons. It is assumed that she made norm-referenced judgements when she was planning the unit, as she would have taken into account what the usual, normal or average group of students of this year group would be able to understand and learn in the unit being planned. It is interpreted that at this time she made decisions about what learning experiences and outcomes were appropriate for the class, based on her knowledge of year 8 (aged 12-13 years) students. The teacher's ipsative and criterion-referenced assessments are now discussed in turn.

Ipsative Interpretations

Ipsative assessment is when a student is assessed against her or his own previous performance and is an important component of formative assessment when a teacher wishes to interact with a student's thinking in order to better mediate learning. Teacher 5's frequent, unprompted, use of the word 'expected' while talking to the researcher was interpreted as a key indicator that she used ipsative interpretation on the assessment information she gathered from the class and from individual students. In order to expect something to occur, or to be surprised, it is necessary to have established a basis for that expectation. It was interpreted that the teacher had previously assessed her students and had arrived at a judgement about their level of knowledge, skills or their attitudes. These previous judgements were what formed the basis for her interpretation of new information. That is, she compared a student or the class with their own previous performances. She used ipsative assessment. For example, when speaking generally of individuals her comments suggested that she expected that particular students would understand new ideas and be able to complete tasks (T5/D10/95b).

She continued this comment by naming several students who had not met her expectations. After another lesson, which included an extensive debate on whether or not substances expand in the cold, she said:

> Well, there was assessment in that it was great to know L didn't let me down and my assessment of her is correct. (T5/D15/95b)

She also spoke of having expectations of the class and she appeared to interpret the information she had gathered on the basis of her prior knowledge of the student or class. Sometimes the new information matched with her prior knowledge in that she found out what she expected. At other times, the new information did not match her expectations and she was surprised by it (T5/D15/95b). She spoke of being surprised that particular students were coping and others weren't:

> ... the surprise comes occasionally when I think so and so will have a problem and I go over there and she's right on task, it's making sense to her. And then some other person whom I think should be fine, I'll just let her get on with it, asks me a question, and says, 'Can you help me because I don't understand this kind of thing'. (So did that happen today?) Yes it did. There ... was a student who I thought would have understood, who wouldn't have had a problem interpreting the map, and she had to clarify it. Over there (she pointed to where the student sat), she needed an extra question to keep pointing in the right direction. A person over here whom I thought might be confused by it, she was actually right onto it, ... she had seen there were two lots of information and she was focusing on the plates which is the one that I really wanted to focus on. And then this other person over here, she was struggling ... So that was a bit of a surprise, I thought that she might have ... (T5/D10/95b)

For teacher 5 to use ipsative assessment, she must have previously formed an impression of a student or the class. She must have had prior knowledge of a student. Formative assessment is a process which is intended to assess and then inform student learning during that learning. Within this process, the teacher's learning is also promoted. Teacher 5 learnt about a student or the class and the effectiveness of her teaching approaches. If it is assumed that the teacher's learning proceeds through conceptual development, this raises the question of how teachers conceptualise their prior knowledge of students. On one occasion, the teacher spoke of using new information to refine the 'picture' she had of a student. This teacher stored her knowledge as a picture of the student:

> ..... no I didn't find anything new about A ... what I did find out was, ... the picture I have about A is a little bit clearer. On the surface of it, she looks like she's going OK and understanding, but when you pin her down, so she in actual fact needs .... (T5/D14/95b)

It appears that the picture teacher 5 had of this student included her impression of the depth of the student's understanding. Whether all teachers store their prior knowledge of students in the form of a 'picture' is unknown, as is whether this picture is linked with a teacher's gut feeling' judgements (Cowie and Bell, 1995).

By talking of being surprised by some students, teacher 5 highlighted the need for on-going formative assessment. She stated that students do not always meet a teacher's expectations and so on-going informal assessment is essential to ensure their learning is maximised in every situation. She did not view the students as being static:

> (So it was a surprise, yes she could. And then others, it was a surprise, no they can't) Imm, that's right. ... that's important because so often we group the children and we have in terms of what we expect. We expect that some children won't have problems, they will follow what we say, because this is what they generally do most of the time. ... but it's really important that we don't do that all the time. That we do go and check on these children. It maybe that the one time when they really need to be clearer on what they understand, that's when they are off the track. And you cannot just assume that because most of the time they are on track that they will always be. (That they will be then) and also I suppose, children who tend to have more difficulties following and understanding, who take more time to develop an understanding, you assume that they'll take more time, every time. ... and there are times when it's quite neat, because in fact they've got onto it. (T5/D10/95b)

Here the teacher highlighted the need for on-going assessment.

In summary, teacher 5 used ipsative assessment as a basis for some of her interpretations of the formative assessment information she gathered. In order to use this form of assessment, she had to have made a prior judgement of the student. Hence, interpreting the formative assessment information she had gathered involved her in a process of conceptual development - conceptual development of her knowledge of her students. She indicated that it was important that she undertook on-going assessment of her students as they did not always meet her expectations.

While teacher 5 used ipsative assessment she also used criterion-referenced assessment.

Criterion-referenced interpretations

Criterion-referenced interpretations in formative assessment involve the teacher comparing an individual's or class's performance on an objective with a predefined set of criteria, which detail the levels at which the objective may be met. Such assessments and interpretations are typically curriculum-referenced. In New Zealand, *The New Zealand Curriculum Framework* (Ministry of Education, 1993a), details the learning outcomes which the Ministry of Education considers should be promoted within New Zealand schools. The *Science in New Zealand Curriculum* document (Ministry of Education, 1993b) elaborates on these for a students' science learning.

Teacher 5 was observed to use both task related criteria and concept-related criteria.

Task-related criteria

Teacher 5 formatively assessed whether the students understood the requirements of the task being used to promote the learning. Such formative assessment was often informal and on-going. Teacher 5 observed and listened to the students while they were engaged in the task. It is interpreted that teacher 5 decided whether the students understood the instructions, were able to use a piece of equipment or were able to complete the task because they lacked the prerequisite understanding or knowledge. Such formative assessment was seen as essential to ensure the students were able to complete the task and for the smooth running of the classroom.

For example, teacher 5 assigned her students six tasks at the beginning of the unit which was observed. She handed out additional material during the unit. On one occasion, she handed out a map of the world with New Zealand and Australia in the centre. The continental crusts, the tectonic plates and the sites of volcanoes were marked on it. She asked the students to identify and colour the plates. During an end-of-lesson discussion she said:

> ... I found out that the map was a little bit confusing for some students, not the ones that perhaps I would have expected it. There were two lots of information on that map. There were the crusts, the oceanic and the continental crusts, and also the plates. What I wanted them to do was to fill in the key. And I don't think I was that clear about that ... by the time I got to giving out instructions, I had given a few too many, ... and I wasn't clear enough on that. The ones who seemed to be confusing it, I knew they had confused it, because they were colouring in the crusts, rather than the plates. .... I talked with them about that and whether they could see the difference. ... It's perfectly obvious to some people, but to others was confusing. Yet if you choose another piece, another map, those who were not confused might be and vice versa, so .... it's a matter of using a variety of things to get your point across, so that there's something there for everybody. (T5/D10/95b).

Here teacher 5 described formatively assessing the students by looking at their work. She interpreted that they were confused because they were not colouring in the map in the expected manner. Her criterion for the successful completion of the task was that the students coloured in the tectonic plates. Through this criterion she gained access to the conceptual criteria for the task. This was that the students could identify the tectonic plates. In this case, the criteria for completing the task appeared to be closely linked to the criteria for the conceptual learning within that task. It is proposed that when the criteria for completing the task are simple, these may provide easy access to evidence of a student's conceptual understanding.

Concept-related criteria

When teachers plan their teaching to promote the understanding of a particular concept, they are able to anticipate some of the criteria students will meet when they are demonstrating their understanding of that concept. Teacher 5 emphasised knowledge and the use of scientific vocabulary as criteria she used to make judgements about the students' science content knowledge. For example:

> J , E and B are very confident. What they think they know, they do know. Their language is appropriate, they use the right words. These are the judgements I make listening to them. (T5/D14/95b)

The use of appropriate vocabulary is one aspect of content. Depth of understanding is another.

Teacher 5 also assessed the students' ability to link the science they were learning in the classroom to their everyday experiences. The data suggested that this ability was also used as a criterion to assess the development of a student's understanding of science concepts. (T5/D12/95b)

In criterion-referenced assessment, it is common to detail four or five levels at which a student may meet an objective. Formative assessment, often of necessity, requires on-the-spot interpretation and action so it is possible that teachers would need to use fewer levels in it (Bachor and Anderson, 1994). For example, teacher 5 spoke of the two levels she used, general and specific knowledge. After questioning the students as a group, she said:

> ... that showed me that, yes there's a lot of knowledge there now and that the next step is perhaps to get children to make sure they are clear on this and start getting a bit more specific. We've got a lot of general information coming through. Now we need to look at some specific things, and I suppose that's why we launched into the worksheets today, to sort of focus on specific things. (T5/D11/95b)

In this instance, teacher 5 used a two level criterion to judge the development of student knowledge.

Other aspects of making interpretations

Teacher 5, like other teachers, was responsible for the progress and learning of the class as well as that of individual students. Once a lesson had started, the need was to assess the effectiveness of the learning activity in terms of the learning it was promoting for the class as a whole. To do this, teacher 5 often interpreted the formative assessment information to form a 'general impression' of the class based on the number of students who were able to answer questions or complete tasks. For example, when evaluating the video she had shown, teacher 5 said:

> ... I just asked again some general questions. ... when you ask general questions usually you can gauge how well the information has gone in by the number of children who respond. ... I think that I use that quite a lot. If I only get one or two people, and I look to see who it is, and if it's my really bright children, whom I know have the good skills, good data gathering skills over a range of ways, then I think, ' OK this is maybe not hitting the middle mark'. ... And I've got to do something more. (T5/D11/95b)

In this instance, she not only noted the number of students who were responding positively but also who those students were.

The teacher involved in this case study also interpreted a student's willingness to offer suggestions during whole class discussions as reflecting the student's positive self assessment of her own progress. For example, she said:

> ... the relevance of what we were talking about to their everybody life.(Was there anything in particular that gave you clues to that some of them might be coming closer to ...?)... She had some input today and she's been quite quiet up until today. ... I think this is a sign. That when you start to make connections, you want to verify this and so you ask. And I think that's what she was doing. I felt that E, she might have been along the line a little bit further. (T5/D14/95b)

In summary, teacher 5 interpreted the formative assessment information she gathered. To do this, she used norm-referenced, criteria-referenced and ipsative interpretations. Of these, criteria-referenced and ipsative assessment were considered most important for formative assessment. The interpretations teacher 5 made formed the basis for her actions, which are discussed in the following section.

3.9 TAKING ACTION

The next aspect of the process of formative assessment worth noting was the action the teacher took. Teacher 5 had the possibility and choice of action once she had gathered and interpreted the formative assessment information. It is proposed that teacher 5 chose when she acted, with whom she acted and how she acted.

Retroactive, interactive and delayed forms of action

Teacher 5 acted retroactively, interactively and in a delayed way. In taking retroactive action, the teacher and the student revisited the concepts or the learning task which was being used to mediate the learning. For example, teacher 5 spoke of doing this when she assessed that the students were having difficulty with identifying and colouring the tectonic plates on a map. She talked with the student revisiting the requirements of the task and the concepts involved:

> There was lots of information ..... I don't think I made it clear ..... The ones who were confusing it, I knew they were because they were colouring in the crusts rather than the plates. I talked with them about that and whether they saw the difference and how they could change it. (T5/D10/95b)

Teacher 5 also used interactive action. That is, she interacted with the student or the class in the moment, on the basis of the understanding they were demonstrating at the time. Teacher 5 often acted immediately, when information was collected during on-going and informal interaction between the teacher and the student. The information collected at this time usually indicated students were having difficulty with a task or concept. For example, when interacting with another student over the map in the example above, she suggested the student get out a simpler map, then she and the student discussed this (T5/D10/95b). In this example, she used other materials to help with the student's learning. At other times said she referred students to other students whom she considered they would help or to books (T5/FN9/95b). For example, when she assessed that the students were confusing the effects of heat and cold, she asked some students to model particles being heated (T5/FN14/95b). Teacher 5 also took immediate action when she assessed that individual students were not meeting their expectations in the areas of personal and social development.

The students in the class involved in this case study appeared to interpret the fact that the teacher was collecting and recording assessment information as an interactive assessment action on her part. On one occasion, the teacher systematically assessed how many tasks the students had completed. She recorded the names of those students who were behind. It is interpreted that the students took this as an assessment action because, when the teacher checked on them the next day, they had made considerable progress with the tasks (T5/D12/95b). The teacher commented that with this particular class the recording of names was sufficient to focus students on their work.

Teacher 5 also chose to defer the action she took - that is she took delayed action. For example, on one occasion she interacted with individual students in the second half of the lesson. At the end of the lesson, she told the researcher that the students were confused over ideas to do with soil, land forms and ore (T5/D10/95b). She acted on this information at the beginning of the next lesson, by leading a discussion on soil and land forms (T5/FN10/95b). She did not consider she had sufficient time to take action at the time when she collected the information.

Teacher 5 frequently stated that she tried not to give knowledge until there was a reasonable level of interest and knowledge within the class. She referred to this as the knowledge 'growing' within the class. She watched, listened and waited for this to happen. On one occasion when she determined there was a general level of knowledge, she handed out more specific photocopied information:

> We've got a lot of general information coming through. Now we need to look at some specific things. I suppose that's why we launched into the worksheets today, to sort of focus on specific things. (T5/D11/95b)

It is proposed that the possibility of this type of delayed action often depends on the time the teacher has available for the unit, the nature of the topic and how the concepts and skills are connected within it. For example, this teacher followed up the concept of the composition of soil on three occasions.

By waiting for knowledge to 'grow' within the class, the teacher was often able to use students to input the information. For example, when students asked about the nature and source of gases, she asked them to research this for homework. Those who had followed this up shared their knowledge the next day (T5/D14/95b). On another occasion, when the students debated whether substances expand or contract in the cold, she facilitated sharing and interaction between the students based on their prior knowledge and experiences of this. Her main contribution was to keep the discussion focused and to draw it to a conclusion.

With whom did the teacher act?

Teacher 5 also had a choice of whom she acted with. Teacher 5 could gather and interpret information from an individual or the class but she might then act with the same student(s) or with others. For example, teacher 5 assessed that individual students had misunderstood the concept of a land form and then discussed this with the class. In this instance, the number of individuals with misunderstandings made it more profitable for her to act with the group rather than with each individually. At other times, she gathered information on the class's understanding and acted with the class. For example, teacher 5 gathered information which confirmed that the majority of the class did not have scientifically acceptable conceptions of the effects of heat on solids. She included all students in the resolution of their confusion through the use of a whole class discussion (T5/FN14/95b). Teacher 5 also gathered information from individuals and then interacted with that individual (T5/D10/95b).

Factors informing the teacher's actions

Two factors were identified as informing the teacher's actions, namely knowledge and experience, and finding out that the students had understood.

Knowledge and experience

How teacher 5 acted appeared to be informed in part by her experience and confidence with the topic and the class. For example, when the class confused the effects of cold on solids, teacher 5's main input was to ask questions which focused the class on the issue of expansion. She also occasionally supported the students' examples with personal anecdotes (T5/FN14/95b). During the end of lesson discussion, teacher 5 stated that she wanted to give her students an opportunity to put forward their ideas on the topic (T5/D14/95b). She said she was confident that they recognised this as a strategy she used. She also stated she considered the concept was one with which they had many everyday experiences and that she was confident that she understood the concept fully. It was significant to the teacher that there was a right and wrong answer. She had prior experience in teaching this concept and was confident that the class would negotiate an understanding that most substance expand with heat:

> What might have been critical at one stage was that they all seemed, as a group, that they were absolutely adamant they were right. At one point I thought to myself ... I thought 'Have I got it wrong?' or 'Have I got it right but what I'm saying is wrong?' So that it was right in my mind but what was coming out of my mouth was wrong. ... then I thought 'No, I've been doing this for such a long time and it's an everyday occurrence'. It isn't one of those scientific theories which gets talked about in science and never again. It is an everyday thing, expansion (T5/D14/95b) (So was it important that it was a right and wrong and it was very clear cut?) Yes because I knew that .... I was pretty confident that the examples they were going to bring were going to fit or they wouldn't. (T5/D15/95b)

In this instance, teacher 5's knowledge of her class, her teaching style, the content and her experience with teaching it all contributed to the action she took in this case. Given her confidence and experience with this topic, her action was to allow the students to resolve the matter for themselves.

Finding out that the student had understood

How teacher 5 acted also appeared to be informed in part by her finding out that the students had understood. Teacher 5 described how she would find out that the students were learning what she expected and then carrying on (as the action taken):

> The general (information) is generally in the form of what I'm expecting to find out and I do find out and this is why ... when nobody else is in the classroom, there's just me in the class, I find it out and I think 'Yes', and I carry on. When somebody asks me specifically about that, I've got to say, perhaps I'm finding out more at that time than I thought. (T5/D11/95b)

During the end of lesson discussions, teacher 5 often spoke of finding out what she expected to find out. As she stated in the last quotation, she was surprised by just how much she had found out during the lesson. On another occasion when she found out that learning was proceeding satisfactorily, she spoke of her mental picture of a student, which was the accumulation of her year's interactions with the student and she updated it when she gathered additional information:

> A , ... I didn't find anything new about A but I did find out was, ... what I found is a bit clearer, the picture I have about A___ is a little bit clearer. ... on the surface of it she looks like she's going Ok and understanding but when you pin her down... (T5/D12/95b)

It is assumed that the teacher used her mental pictures of individual students as the basis of her ipsative interpretation of the information she collected and any individualised action she took with a student.

Teacher 5 had also formed general impressions of the class, which she used to inform her actions. The maintenance and refinement of these pictures was therefore an important action as a part of formative assessment which was responsive to either the class or individual needs. For example, after the class watched a video, the teacher asked some general questions. In order to decide what to do next, she observed how many and who in the class could answer her questions. She then continued by stating that she considered the video was meeting the needs of most of the students, but that if it hadn't been, she would have supplemented this activity with other material (T5/D11/95b). When teacher 5 spoke of being surprised, expecting students to do something, or when she talked of the need to continually assessing students, it is interpreted that she was aware of and continuously updating mental picture of the students (T5/D 10,11,12,14/95b).

Evaluating the formative assessment action

Teacher 5 sometimes evaluated her formative assessment actions. For example, after a discussion on what constituted a landform, during which time each student had contributed their ideas, the teacher asked the class if they understood. The class replied in unison that they did. In this instance, she evaluated her action by asking the class. However, perhaps in recognition of the limitations of this strategy, she stated to the researcher during the end-of-lesson interview that she doubted every student had grasped this idea and that she would follow it up again later. Teacher 5 also asked individual students if they had understood. For example, a student asked a question and she answered. She then asked the student if her reply was appropriate:

> I asked her, 'Is that what you meant?' and she said 'Sort of'. (T5/D14/95b)

Teacher 5 sometimes moved through more than one cycle of the formative assessment process to produce the intended change in student or class understanding. For example, when a student told her that she didn't understand a map, the teacher initially asked the student to ask her a question which would help. When the student was unable to do this, they took out a simpler map and looked at it together.

In summary, the substance and form of the formative assessment action, which was taken by this teacher, was complex. It was influenced by how she gathered and interpreted the information which precipitated the action. It was informed by factors associated with her knowledge of and experience with teaching the subject and with her knowledge of and experience with the student or class. Within this framework, teacher 5 chose when, with whom and how she acted.

3.10 SUMMARY

The aspects involved in the process of formative assessment have each been explored separately, namely the setting, the teachers' views of teaching, learning and assessment, and the purposes for doing it. These aspects are interrelated and interdependent in that each aspect has consequences for the next. By considering each aspect in the process, it is possible to gain an insight into their individual complexities. However, in order to appreciate the reality of a classroom and the significance of their interaction, it is necessary to reflect on the complete process of formative assessment as it affects the teacher and students. Three formative assessment episodes or cameos from the classroom observations of teacher 5 will now be presented.

3.11 THREE CAMEOS OF FORMATIVE ASSESSMENT

Formative assessment in this classroom was characterised by its integration into the teaching/learning process, a high degree of student choice and the teacher's assessment of students' social and personal development as integral to and supportive of their learning of science. Teacher and student formative assessment actions were supported by the expectation that ideas would be shared and respected, and that there was a weak boundary constructed by the activities and teacher feedback, between school science and student's everyday experiences.

Assessment in the case study classroom is illustrated through three cameos. The cameos are considered to be episodes, where an episode is defined as all that happens from the time when the teacher started collecting the assessment information to when she or he had finished carrying out and evaluating her or his action.

Cameo: Soil composition

The time when the class discussed soil composition is presented as a cameo to illustrate the integration between formative assessment, teaching and learning; the influence of the weak boundary that had been constructed between school science and the students' and teacher's everyday experiences; and the teacher's waiting for the knowledge of the class to 'grow'.

The teacher's response to finding a number of students had an alternative understanding of the composition of soil through interaction during small group work, was to pose a question to the whole class at the beginning of the next lesson. The episode was field noted as:

> The teacher asked the class: 'Why is the top layer of soil darker?' Students responded by suggesting the soil became sunburnt in the same way people do, they described burnt cakes, the colour of dry areas, the colour of compost, the top layer of the soil when they were on camp (the class had gone on a class camp earlier in the year) and the colour of damp soil. Twenty students contributed anecdotes from their experiences to this discussion. The teacher also contributed anecdotes on the use of compost in her garden, going on camp and the colour of the soil in the school quad. It was agreed that the top layer of soil was usually darker but no consensus explanation of why emerged. The teacher concluded the discussion by stating 'I think we made progress on that question. Are there any questions?'. No one replied. (T5/FN10/95b).

This episode illustrates the teacher's typical response to finding out students' held alternative conceptions. The questions she posed elicited student ideas while simultaneously providing feedback.

The episode was of interest because the patterns of discussion established during the whole class sharing times at the beginning of each lesson, were crucial to the viability of the teacher's action. It was characteristic in that twenty students drew on their experiences and contributed anecdotes. The students' immediate contribution of a wide range of ideas that suggested they viewed school science as linked to their everyday experiences. The teacher's own contributions during this episode also supported this linking. Her action was consistent with the value she said she placed on students linking what they were learning with their everyday lives (T5/D18/95b).

The teacher's response to the students' uncertainty about soil composition was also characteristic in that she introduced this as a topic for discussion on two other occasions. On each occasion, she encouraged students to contribute ideas and to seek out more information for themselves. She only input information herself on the last occasion. The development of the students' ability to conduct research (a personal skill) and share the results of their research with others (a social skill) was one of the teacher's long term goals for student learning (T5/D11/95b). She used this strategy because she considered sharing their ideas developed understanding (T5/D10/95b).

Hence, the teacher's assessment of her students' personal and social development was also linked with her assessment of their science development.

The teacher's delay in inputting information was derived from the view that she needed to wait for the development of collective knowledge (T5/D8/95b). She viewed the class as an 'organism' whose knowledge and interest 'grew' and monitoring the development of this knowledge was a particular focus during the discussions at the beginning of each lesson. She explained this action when analysing a discussion of rock types:

> This morning we tried to talk on rock types but there was not sufficient knowledge to
> sustain a conversation therefore I will seek to encourage knowledge in this areas and
> wait until the collective knowledge and interest is great enough before we proceed. ...
> In this case we will move out sideways and wrap back (T5/D5/95b))

As a number of students reported to the class on ideas they had explored, for example the nature of alluvial soil (T5/FN4/95b), composition and use of natural gas (T5/FN6/95b), it appeared this approach fostered the view that students were able to contribute to each other's learning.

Another example of the use of planned formative assessment occurred when the teacher noticed that some students were uncertain as to what counted as a land form. This next cameo illustrates the variation in the actions the teacher took.

Cameo: Land forms

One of the set tasks for the unit required the students to draw a map of the landforms in the local area. The teacher noticed some of the students were confused as to what constituted a landform, for example, were lakes and trees landforms? The next day, during the whole class discussion time, she asked each student in the class to suggest a land form and identified whether or not the student was correct. She elaborated on some local examples, such as Hinuera stone. Next, she checked whether the students could identify counter examples, for example, she asked, 'Are forests a land form?'.

(This had come up as an example during a previous lesson.) (T5/FN9,10/95b). She concluded the session by asking, 'Any questions?'. The students assured her they understood the idea.

This episode is of interest because the teacher's actions contrast with those in the previous Cameo: Soil Composition. The topic involved understanding a definition and the teacher required every student to disclose their thinking. She used her authority to legitimate some students' examples as land forms. In addition, she assessed the effectiveness of her action by asking the students to differentiate between examples and counter examples of landforms and concluded the lesson by providing an opportunity for students to seek further help.

Another episode that illustrated the complex and situated nature of formative assessment in this classroom occurred when a student presented her investigation of weathering to the class.

Cameo: Expanding and Contracting

The students were required to present the results of their exploration two questions to the class as part of their summative assessment. They were able to choose when and what they presented. This cameo began as a student was presenting her findings about the effect of freezing a piece of wet chalk. The researcher was absent from the classroom when the episode began. The episode was field noted from the time the teacher asked for a show of hands to confirm whether the students thought metals expanded or contracted when heated:

Nearly all students indicated they thought substances expand in the cold. A student explained this by recalling what happened to the metal teeth on bridges.

The teacher asked eight students to move to the corner and to jump up and down. She asked them if they needed more room to move when they did this and what had happened to their temperature. A discussion followed as to whether this showed substances expand when they are hotter or whether it showed that particles move around more when they are heated.

The teacher encouraged the students to contribute their ideas and experiences and to make sense of all contributions. A student recalled a ball and loop experiment the class had performed the year before. She explained this as the ring contracting when it was heated. One student suggested that telephone wires sag in the heat. Another said she had been in Christchurch when it was really cold and the wires were sagging then.

The teacher asked for another vote on the issue. All but four of the students voted that substances expand in the cold. The class suggested their previous teacher be asked what happened.

By coincidence their previous teacher arrived and the case study teacher explained the situation to her. The students' previous teacher was obviously surprised at their views. She left.

The discussion continued. One student suggested that it was the water that had expanded not the rock. More students contributed evidence of solids contracting when they cooled. They stated cakes contract as they cool, hair is longer when it is wet, sultanas shrivel as they are dried. The teacher focused the students on the question. More students contributed explanations which suggested the water expanded when the

> rock was frozen. Other students explained that the ball and ring experiment as the ball expanding when heated.
>
> The teacher concluded the lesson by asking the students what they thought. All but three students indicated they considered cold usually caused solids to contract. (T5/FN15/95b)

This episode was of interest because it illustrated the divergent focus of the teacher's assessment, that is, she responded to student ideas even though the notion of expansion and contraction was not part of her planned unit. The topic of expanding and contracting was not a focus of the unit but Teacher 5 assessed that a student and then a large number of students had scientifically unacceptable understandings. Her action of encouraging the students to share and make sense of their experiences was made possible in part because she had the autonomy to extend the lesson to give time for debate. This action drew on and utilised the already established social norms that students would share their ideas and experiences and respect the ideas of others. These norms were such that three students were prepared to disagree with the consensus opinion at the end of the lesson.

The students' immediate contribution of their own experiences was consistent with the weak distinction the teacher had maintained between science and the students' everyday experiences. Presumably, it was also a reflection of the topic. Interestingly, although the students used empirical evidence (albeit recalled) to persuade each other, they deferred to the authority of their previous teacher when she arrived. During this episode, the teacher's asking the students to share ideas and reach a consensus, along with her action of asking students to model expanding, construed them as thoughtful and having ideas and experiences to offer. It served to represent students as meaning makers, science as linked to their experience, scientific explanations as making sense and consistent with empirical evidence.

The teacher's view of the episode

The teacher described the episode as one in which she had encouraged the class to deconstruct the idea of expansion and contraction:

> Is that what you call deconstructing? ..... Breaking it down and finding out what the bits are. What bits have we got? I think the bits were all there but they just had them in the wrong order. So we had to take the concept apart and see what is was we were trying to find out. (T5/D15/95b)

And then to reach a consensus:

> I'd given some sort of clue as to what we were going to do. We were going to have to agree on something and it was either this or that. Nothing in between. (T5/D15/95b)

Confidence in her understanding of the science, pedagogical skills and knowledge of the students played an important role in her spontaneous action. She was confident she understood the science as 'there is no doubt about .. what heat will do to metals'. She was confident, based on her experience of teaching the topic (her pedagogical content knowledge), that the students would have everyday experiences to share:

> I know there are lots of really good examples and I felt sure we could bring those example to light and the kids would be convinced.   ... It is an everyday thing, expansion. (T5/D15/95b)

Her choice of action was influenced by her confidence the class had the skills to reach a consensus and would recognise she intended them to do so:

> I have confidence that I and they have developed certain skills and patterns. I think they recognise this technique of discussing around. I don't say 'No' to someone. I say 'Ummm' and I go onto the next person. That indicates to children 'Well that person might have had an idea, but it was a bit deep, it was a bit hidden, or they weren't on the right track'. But who knows. So I go onto the next person to see if they can give something. It maybe critical, in that the technique may not be an option if you don't know your class. It is something you have got to develop. (T5/D15/95b)

And, that she had the skills to help the students 'agree on something'. Even so, her confidence in her understanding and her communication skills wavered during the discussion when the class had seemed 'absolutely adamant they were right'. This prompted her to question her own understanding and communication skills:

> 'Have I got it wrong?' or 'Have I got it right but what I'm saying is wrong?' at this time. So that it was right in my mind but what was coming out of my mouth was wrong. (T5/D15/95b)

Teacher 5 indicated she had been surprised by the thinking of some of the students:

> Well, there was assessment in that it was great to know L didn't let me down and my assessment of her is correct. I was a bit stunned at J being on the fence and E begin on totally the wrong track. ... (T5/D15/95b)

She reported she was satisfied that all but three students understood the ideas of expanding and contracting at the end of the lesson.

Student perceptions of the episode

This episode was discussed with two students, who entered the room while the researcher was talking with the teacher immediately after the lesson and with six students a day later. The two students said their knowledge of the teacher's usual actions and reactions had confused them during the discussion. They associated sustained teacher questioning with them not being 'on the right track'. They told the teacher:

S51	I thought it was a bit strange because you kept asking questions.
S59	And you didn't go 'Yeah, that's right' or anything.
S51	This is strange, but I'm sure I'm right.
T5	Yes, is that a technique that I use, that when people aren't on the right track, I keep asking questions. But when people are on the right track...
S59	You don't say much.
S51	You don't go like this, 'Ooh right, and what do you think?'.
S59	And like if it's right, you'll say 'Aahhh'. (T5/D15/95b)

Other students were also interviewed at the end of the unit of work. They were ambivalent about whether the discussion had contributed to their understanding. Four said they had become 'mixed up'. Their view was illustrated by the student who said:

> ... well I got a bit mixed up with that hot and cold thing. Which got bigger? I thought that the power lines actually drooped when it was cold, but it's the other way around. And I didn't actually realise that liquid and um, solids are two different reactions. (S54/I/95b)

However, they stated it was important to share ideas and that eventually the discussion had resolved their confusion.

Student comments (and reactions) indicated they attributed their teachers with the authority to legitimate answers as right. They would have preferred the teacher to exercise her authority sooner as one student explained:

> I think I wouldn't have been confused if X [the teacher] had said like, told us the answer. Then I wouldn't have got confused. (S54/I/95b).

Teacher body language was reported as able to legitimate ideas. One student reported it could have influenced the whole discussion, she said:

> If T's face had gone [the student raised her eyebrows]..... as we'd started the discussion we would have all changed our minds straight away. But she didn't. (S51/I/95b).

Their previous teacher's reaction to their assertion solids expand with the cold had been crucial in 'convincing' them this was not so. A representative comment was:

> At the start when I said that the cold makes it expand it kind-of felt, sounded funny. I thought 'Oh', but I just kept saying it until I convinced myself that I was right. I then unconvinced myself when I saw Mrs X's face. (S54/I/95b)

In this case, the student ignored her intuition the answer was wrong and was persuaded by the consensus view and then by the teacher as an authority on science.

The students indicated that some students were attributed with an authority commensurate with that of a teacher. In this episode they claimed that if, 'L', a student they (and the teacher) considered to be bright, had given the correct answer early in the discussion 'everyone would have agreed with her' and the debate would not have ensued (S51, 59/I/95b).

Recalled empirical evidence was also reported as influential. One student said a peer recalling snow on the telephone wires near Christchurch (a city in the south island of New Zealand) had been particularly persuasive. She explained:

> B said 'I remember when I was in the South Island and it snowed heavily and the power lines were really down low. It was really cold'. That swayed or slightly convinced me because I thought, 'No, she wouldn't forget it, if they were down there, they were down there. It isn't something you make up or forget' (S51/I/95b).

They also identified the ball and ring experiment from the previous year as influential although it appeared that they had 'remembered the experiment quite clearly' but forgotten 'which way around it was' (S51/I/95b).

The students were emphatic they needed to 'find out whether the ideas are right or not'. They identified teachers and text and empirical evidence as having the authority - being trustworthy enough - to do this as was illustrated by two students who said:

S56   You have to eventually either read a book or do an experiment to find out.
         Because otherwise you just have a whole lot of different ideas from
         different people.

S55   And you don't know what was right.   (S55, 56/I/95b)

Another point of interest in this episode was that the teacher and students spoke of the discussion in terms of their ideas moving along a path or track. The teacher reported her intention had been to 'find out where this path may led and we'll make it come back to where we need it to be'. She had been surprised E was 'on totally the wrong track'. It had been critical that she understood what the students were thinking and 'where we have got to get to' - the scientific explanation of expanding. The challenge had been to 'figure out a pathway there'. She noted the pathway had 'come right back to where we started but rather than to see expansion as with cold seeing it with water'. This notion of learning as movement along a track was also used by the students. A representative student comment was:

we had a discussion and we kind of, she said that, like put up your hand for ideas and if we went off track she'd make sure that before we left the discussion we were on track, sort of thing. (S51/I/95b)

This suggests both teacher and students had a sense of purpose or direction for their engagement in the discussion.

3.12 DISCUSSION AND SUMMARY

Formative assessment in this case study classroom was shaped by the learning activities and assessment practices utilised by Teacher 5, the time of the year, the topic of the unit and the nature of the classroom. Lessons followed a set format of whole class discussion and individual work on set tasks. During these activities her assessment focus was on the students' social, personal and science development and included, for example, the ability to listen and share ideas, to manage the time and undertake research and student understanding of the curriculum she intended them to learn and the understandings and interests they were developing. Teacher 5's assessment practices were characterised by an integration between teaching and assessing. She formatively assessed and responded to student learning when she interacted with students while they were working on the set tasks. She followed up some these informal assessments with a planned whole class discussion stimulated by posing a question at the beginning of a lesson. That is, her planned and interactive assessment interacted and informed each other.

The unhurried nature of the teacher's management of the unit activities, made possible by her autonomy and the institutional setting, seemed to be important. The teacher revisited the same idea on a number of occasions, she encouraged students to think about and find out about ideas. These actions construed understanding as something which takes time, students as members of a community of learners who were able to contribute to development of collective understanding and coherent explanations.

The weak boundary the whole class discussions constructed between school science and student's everyday knowledge and experiences, provided the teacher with a rich source of robust information on student interests and the links they were making. The discussions provided students with feedback on what counted as school science and the standing of their own ideas

Assessment was shaped by, as well as shaping, teacher and student expectations that ideas would be shared and respected. Teacher 5 utilised students' ability to discuss ideas to encourage them to contribute ideas and experiences to develop consensual explanations to ideas when they were confused or held scientifically unacceptable explanations. These expectations supported student disclosure to the extent that three students were prepared to publicly state they disagreed with the consensus view of the effect of cold on solids. In this way, the teacher's assessment practices construed school science as the coherent explanation of empirical evidence with the students being seen as able to develop these explanations through debate and negotiation. The students indicated they considered themselves in this way, as one student explained:

> We were talking in the car, because we're in a car pool, and [S51] said that it's a proven fact that children learn more if they find out for themselves. (S57/I/95b).

The students were active in the formative assessment process. They assessed their own and each other's book work and sought help and advice about ideas and how to do tasks from peers and the teacher. These actions were consistent with the teacher's view that by the end of the year students should share the responsibility for their assessment. However, the students' responses in the discussion about the effect of cold on solids indicated this had been only partially successful. Although the teacher guided them towards a consensus view it seemed that they were persuaded by the authority of their previous teacher and peers they considered knowledgeable, to the extent they reconceptualised practical results in contradiction to their own sense of what might happen.

A striking feature of this case study was the similarity between the students' and the teacher's perception of what was learned. The teacher considered the students were beginning to appreciate the links between school science and their lives and develop more 'specific' understandings of geological phenomena (T5/D12/95b). The seven interviewed students agreed, one student said that whereas previously she known what land forms were now she understood how they were formed (S51/I/95b). Another said she had known about rocks and weathering but not the different types of rocks (S54/I/95b). A third said she had known there were tectonic plates but not what they were (S53/I/95b). The students said they had found out about things they had previously taken-for-granted and now found fascinating (S51-57/I/95b).

The teacher's and students' views of learning and assessment were also similar. The teacher's comments indicated she viewed learning as movement along a path or track towards a predetermined destination. It seemed she tolerated divergent pathways for student learning and that her assessment was aimed at guiding the students towards her destination (the science) over time. A metaphor for learning as movement along a path or track was also used by the students (S51/I/95b). Thus it appeared the teacher and students viewed school learning as purposeful but not strictly controlled by the teacher, so that the students were active participants in learning.

In summary, this chapter documents a case study of the formative assessment done by a teacher and her students in a unit of work on earth sciences. The case study illustrates key findings that are also collectively found in the other case studies. These are that formative assessment is a highly contextualised activity, that is, exactly what is done by way of formative assessment is influenced markedly by the context. Formative assessment is a purposeful, intentional activity. It involves verbal and non-verbal interactions between the teacher and students and the eliciting and interpreting of information and taking action. These aspects of formative assessment are elaborated on in the next two chapters on the characteristics of formative assessment and a model, drawing on data from all eight case studies. Further examples of formative assessment are given in chapter 6.

CHAPTER 4

THE CHARACTERISTICS OF FORMATIVE ASSESSMENT

A summary of the characteristics of formative assessment was made from the data from the eight case studies, including the interviews with teachers and students, the classroom observations, and the discussions on the teacher development days. The ten characteristics of formative assessment identified were responsiveness; the sources of evidence; student disclosure; a tacit process; using professional knowledge and experiences; an integral part of teaching and learning; who is doing the formative assessment; the purposes for formative assessment; the contextualised nature of the process; and the dilemmas (Bell and Cowie, 1997, p. 279). Each of these will be discussed in turn.

4.1 RESPONSIVENESS

The essence of formative assessment in the definitions cited earlier was the component of action or responsiveness of the teacher and students to the assessment information gathered or elicited. The different aspects of responsiveness discussed by the teachers were:

Formative assessment is responsive in that it is on-going and progressive

The teachers involved in the research commented that they felt that formative assessment was characterised by its on-going, dynamic and progressive nature. They commented on the responsiveness:

> If you do something to find out where they (the students) are at, and then you do something from that to change your teaching or what you are doing, then its formative (assessment) ... (TD5/96/14.13) [See the appendix for an explanation of the data codes].

Comments were made that formative assessment was not tied to a specific learning pathway and that the process was flexible and responsive:

> A lot of the time you start off on one tack, and you think, no that didn't work so I'll try another tack, as so its self-assessment (of our teaching) as you go along. (TD4/95/11.41)

Formative assessment was seen as an on-going, everyday event:

> Without formative assessment, teachers do not function effectively. So its your on-going, day-by-day, every-day assessment (TD4/95/11.44)

Formative assessment is responsive in that it can be informal

The teachers referred to assessment as both formal and informal. In saying this, they were usually referring to whether the information gathered was recorded and reported in some way or whether it was used in the classroom activities, without a written record being made. Formative assessment tended to be informal, with no written record of the information gathered. The information was used in the teaching and learning in the classroom and to build up a picture of the student learning by the teacher. For example:

> It may just be how much concrete we set it in .. I can go into a classroom and give the kids a spot ten-question-test because I think they ... just need to do that, to refocus them a bit... I dont record it anywhere, they'll do it in the back of their books. (We) mark it, and I say who got.. this, who got that, 'Thats fine', and we carry on. And that's not set in any concrete at all. (TD4/95/11.74)

Formative assessment is responsive in that it is interactive

The teachers stated that formative assessment was interactive. That is, the information gathered was used in the interactions between teacher and student during the teaching and learning. For example:

> A lot of people haven't been aware .. that assessment can be done at other times .. a lot of teachers .. have just tended to assess students at the end of units and have really not been a part of that interactive process (TD4/95/11.69)

Formative assessment is responsive in that it can be unplanned as well as planned.

The comments by the teachers suggested that at times they planned to do formative assessment but at other times they did unplanned assessments. A planned formative assessment was often used at the beginning of the unit, for example, the eliciting of students' prior knowledge before the teaching of the unit started. A planned formative assessment could also be used to start the formative assessment process within a lesson, for example, a quick ten question spot test at the start of a lesson to find out if the students had understood the ideas introduced in the previous lesson.

The unplanned formative assessments arose from the students' responses, which often could not be predicted and planned for in advance. For example, in taking into account the student view that substances expand on cooling, teacher 5 responded by undertaking some unplanned formative assessment. The words 'unanticipated' and 'incidental' were also used in this context. For example:

> I find that certainly toward the end of the year, children will ask these sorts of questions and so I planned for this to happen, but I never know what they are going to ask (TD5/96/15.4)

The teachers commented that they planned or were prepared for the unplanned, for example:

Planned or unplanned ...Yes, sure you get the kids set up. You don't know what you're going to get. And that's the unplanned part. What comes back from the children. But you get them set up in the first place. ...So you plan the opportunity, but don't necessarily plan the (response). The lesson was planned, this is what they were going to do. But the unplanned part was, oh ...But that the most exciting teaching, when you sort of go tangent-wise. ...I know in that, I've ended up calling it planned and unplanned. I've now gone and changed it to planned and incidental, which just sort of ... cause unplanned makes it sound like you don't know what you're doing, but ... It is the planned opportunity, but there's also that stuff that just opportunistic or spontaneous or some other word that I don't ...Unanticipated. ...Ah, that's better. ...Call it anticipated and unanticipated. (TD5/96/14.16; TD5/96/14.17)

You have planned for the unplanned. I mean, you've left that opportunity for all those incidental things that occur. (TD5/96/15.27)

Formative assessment is responsive in that it can be proactive or reactive

The terms proactive and reactive were also used to indicate the notion of responsiveness inherent in formative assessment. That is, the teacher could be proactive in deliberately seeking formative assessment information from students or reactive, when they undertook formative assessment in response to other information they had gathered about the students' learning. For example:

I thought, it could be proactive where you actually go out and you seek, um, specific times throughout a lesson to actually do the formative assessment. Or it could be reactive. I find that a lot of teaching is, a great percentage is, reactive teaching .(TD5/96/15.12) ... (for example, a) crisis, where students, for some reason, it may be that they are off task, or not prepared or I just them around here, inattentive at listening, or they're all dependent on being followers, lack of ideas or just lost. Crisis point where formative assessment comes in, you have to sort of step in there and take a real lead. (TD5/96/15.14) ... And then, there was a refocussing because you get students who tend to go off track. So by asking questions, on a fairly informal basis, you find out that this kid is way off track and really not going to achieve the objectives that I had planned for the unit. So then you have to get them to refocus again. Like exploring alternative methods and backtrack, taking them back. (TD5/96/15.15)

Formative assessment is responsive in that it involves responding with individuals and with the whole class

The teachers described the way they moved back and forth between the whole class, a group and an individual in their interactions as a result of gathering formative assessment information. For example:

But there is an interesting issue that's coming up for us, and that is that the interplay between the child or the student and the class and how information about the general class feeds into what we do with the student and how information we find out about from a particular student can then feed back into the whole class. That's where the interaction between those two.... Looking at the class or looking at the child are all related. (TD5/96/14.26)

Formative assessment involves uncertainty and risk taking

As the formative assessment done by the teachers was often unplanned and responsive, it involved uncertainties and taking risks. Formative assessment involved the teacher finding out and responding to the diverse views of students; it had indeterminate outcomes; it could not be planned in detail before the lesson; the effects of the required actions were not usually known beforehand; and usually it required the teacher to take action in the busy-ness of the classroom. Their confidence in their professional knowledge and skills was seen by the teachers to influence the degree of risk and uncertainty taken.

Formative assessment has degrees of responsiveness

The teachers said that they had to manage the degree of responsiveness when doing formative assessment. They were aware that they had to manage the behaviour and learning of the whole class as well as that of individuals. They also had to manage attending to the students investigating their own interests and ideas, and to the students learning what was listed in the curriculum. In both these situations, responding to one aspect, meant that they could not respond to the other. They could not always be as responsive to a situation as they wished or were able to.

In summary, the teachers felt that a characteristic of formative assessment was that it was responsive. This responsiveness was discussed in terms of formative assessment being on-going, dynamic and progressive, informal, interactive, unplanned as well as planned, reactive as well as proactive, with the class, group or individual, involving risk and uncertainty, and managing the degree of responsiveness.

4.2 THE SOURCES OF INFORMATION AND EVIDENCE

The second characteristic of formative assessment was the sources of information and evidence. Formative assessment, like summative assessment, may use student written or oral work. But the teachers commented that formative assessment relies on nonverbal as well as verbal information, for example:

> (teachers will be) ... observing kids, in terms of ... facial expressions, body language, listening, talking, writing, .. (TD4/95/12.7)

The sources of formative assessment information for the teachers included the teachers' observations of the students working, for example, in practical activities; the teachers reading student written work in their books, posters, charts, and notes; and the teachers listening to students' speech, including their existing ideas, questions and concerns, and the new understandings they were developing. The teachers set up different learning situations to provide the opportunities for this information to be gathered or elicited. For example, the teachers organised practical and investigative work, brainstorming, spot tests, students recording their before-views, library projects, watching a video, whole class discussions and student self-assessment activities. There was acknowledgment that different learning situations enabled different formative assessment information to be elicited.

An important aspect of this characteristic was that of student disclosure. Student disclosure refers to the information disclosed by students which can be used in the assessment process. Without this disclosure no assessment could happen.

4.3 STUDENT DISCLOSURE

A third characteristic of formative assessment was student disclosure (Cowie, 2000). Disclosure was a crucial aspect of formative assessment that was highlighted in student comments and actions. Disclosure relates to the extent to which a task or activity produces evidence of student performance or thinking. In the classrooms, the teachers used tasks and strategies to elicit student ideas and students voluntarily disclosed their ideas by asking questions and discussing their ideas. Students' perceptions of the disclosure, included teacher assessment strategies, the relationship between teachers' rights and disclosure, disclosure as a source of potential harm and trust as mediating the disclosure.

Data from three sources was collated to map out the breadth of student views. The sources were: (i) individual student interviews with 31 students during phase 1 of the study, (ii) end-of-lesson and end-of-unit discussions with 41 students in phase 2 and (iii) the researcher's participant observations during phase 2. The student interview data from phase 1 may be distinguished from the more contextualised end-of-unit and end-of-lesson discussions in phase two through the coding: data from phase 1 is coded (Sxx/I/95a) and data from phase 2 is coded (Sxx/I/95b or 96). Information on the phases of the research was given in chapter 1, p. 9.

Disclosure and teacher assessment tasks and strategies

In both phases of the study, students critiqued tests, whole class discussions, self assessment and teachers looking at their books as restricting and/or causing them to limit the disclosure of their thinking because of a range of cognitive, affective, social and relational reasons. They indicated one-to-one or small group interaction minimised the negative effects of many of these factors.

It was also suggested that various cognitive, affective and social factors limited the disclosure provided by whole class discussions, observations, self assessment and looking at student books. In whole class discussions, for instance, the social factors of audience and anticipated audience response were identified by many of the phase 2 students as the reason they were reluctant to ask or answer questions. An inappropriate question could, they considered, result in teacher and peer responses that made them feel 'embarrassed' and lead their peers and teachers consider them 'stupid' or 'slow to understand'. Two students explained the class response:

> S54   If the majority of the class do know what they are doing and you don't
>   then it is really hard because it is like 'Ohhhh (sighs), I have to explain
>   it again'.

S95   You feel a lot dumber

S94   And all the other students look at you and you are going (shrinking down in her seat) (SG91/L7/ 96)

The social organisation of self-assessment was identified by some students as impacting on disclosure. In phase 1, two students queried whether self-assessment was a genuine activity. They asserted that they were being asked to record for the teacher what they already knew:

> Well I find them a bit boring because I know that the teacher wants to find out ... about what you've done and how you rate yourself, but it's really just ... telling yourself what you already know. (S51/I/95a)

It seemed that when students were asked to disclose their self-assessment to the teacher, self-assessment, like tests, became a strategy teachers used to elicit information for their own purposes. Moreover, six students said the teacher reaction they anticipated influenced their student self-assessment grades. As one student explained, when speaking about summative self-assessment:

> For self-evaluations I have this little system, and when it says put your mark and it says put your teachers mark, you put what you think is a little bit lower than you think is fair and then when the teacher comes along and you see that she's put ... if you put like, B- and she put A or A- or something like that, you say, oh yay she thought I was worth an A. But if you put something really high and she put something really low, you go away thinking ... hmmm. (S53/I/95a)

This quote suggested that the student tried to out-guess the teacher. This is confirmed by the advice she gave a fellow student that when self-assessing teachers liked students to state what they had done well and then add 'but' and describe what they could improve (S54/I/95b). Self-assessment was observed and discussed with the students of Teacher 2 (Bell and Cowie, 1997, p. 154). They indicated that the need to disclose their assessment to peers and the teacher undermined the fidelity of the recorded information.

The need to maintain their relationship with the teacher was the reason fourteen students in phase 2 gave for the limited validity of teacher observation. They claimed students 'worked' and/ or pretended to be able to do an activity when a teacher was observing them during written work. One student explained:

> Some kids just sit there and they struggle with these questions and the teacher just thinks we're doing OK, cause they act like it. I know heaps of kids ... they pretend (S53/I/95b)

They also claimed they 'pretended' to be listening and understanding during whole class activities. One student explained their actions thus:

> And the teacher can't tell [if we understand] because some of us just sit like this {sitting up and paying attention}... even if we do understand it. ... Sometimes they can't tell just by looking at us ... if we understand it (SG72/L9/96)

Student book work was said to provide little information on individual thinking by six students from the class of Teacher 5. They commented that, as they worked together, their written class work reflected the group view. Altogether, fourteen students in phase 1 questioned why teachers did not assess group understanding, given they often encouraged students to work as groups.

In contrast to other strategies, the students in phase 2 asserted that talking with a teacher individually or from within a small group minimised the negative affective, social and cognitive factors they experienced with other assessment strategies. They considered individual interaction was central to effective teacher formative assessment. One student explained:

> Sometimes she has got to come and talk to you individually because ... if they just say 'Does everyone understand?' , you are going to feel like an idiot saying 'No, I don't'. (SG71/L7/96)

They claimed that when they talked with a teacher they could clarify what she or he wanted to know. Their comments suggested the social consequences of not understanding were minimised when the audience was small - typically the teacher and a group of friends. For example, one student said:

> If someone hadn't understood it, when they were actually doing it [an experiment or other small group task] they'd speak up but when we are just sitting there listening [in the whole class discussion situation] ...you don't really. You can't tell. (SG92/L10/96)

The students considered that their teachers provided them with more useful feedback during one-to-one interaction because they were more explicit about what they did not understand. One group explained that when the teacher 'comes around' they were prepared to ask her 'to re-explain it, just to you personally or to the group'. Three or four students (a sixth of the class) were observed to approach their teachers as soon as small group work commenced. They also talked with the teacher when he or she approached their work space.

In review, the students' critiqued the teacher assessment tasks and strategies as sometimes limiting their disclosure as sought by teachers to gather information on their thinking. These limitations are important given that students need to disclose their ideas before teachers can move through the assessment cycle. Furthermore, student comments highlighted the cognitive, affective, social and relational effects that assessment tasks and strategies have on students, even before the teacher acts to provide feedback.

The relationship between teachers' rights and disclosure

While the students critiqued teacher assessment tasks and strategies, they acceded to teacher requests to participate in them. Only three students indicated that they were unwilling to share their thoughts during whole class discussion, and they acceded to a repeated request to answer a question, thereby highlighting their acceptance of the teacher's rights to require them to disclose their ideas. Teacher and student expectation and acceptance that this was so, was a significant contributor to disclosure and hence formative assessment in the classrooms.

The pervasive and taken-for-granted nature of teachers' rights to require students to disclose their ideas was illustrated by their looking at student books. Student books belonged to the students but teachers looked at them as a matter of course in all ten classrooms. The students described this action as problematic. They were concerned the teacher might see and judge their work when it was 'half finished'. One student explained her feelings thus:

> Well, it's kind of nerve wracking. Cause she's looking at your shoulder, and you're going 'Oh, no, she's reading this. Oh no, it's wrong. It's wrong, I'm way off. Oh, no, oh no.' And when she goes away you can go 'Yes'. ... It's like she's looking at your work when it's just half finished. ... She's not seeing it when it's finished. (S53/I/ 95b)

The student noted this right differentiated between teachers and student - she was not able to see her teacher's work half finished.

Teachers looking at a student's book was made more problematic by teachers' tendency to do so while standing behind the student. The students particularly disliked this because they were aware other students were able to see the teacher's reactions to their work and they were not. They described the practice as 'scary' (SG71/L12/96) and, in the words of one student, as making students feel 'little like a fly on the end of a pin' (SG91/MC/96). It described as 'rude' by the ten students who pointed out it was usually unacceptable. In the words of one boy:

> That's rude. My dad, he doesn't like it when he is reading the newspaper (SG81/L5/96)

Although the students did not like teachers looking at their books the twelve students who were asked said the researcher's suggestion that teachers ask to see their books was 'silly'. They considered teachers were entitled to see their work. One student said her teacher would demand to see her book if she tried to withhold it:

> You can't really say, 'No, you're not allowed to look at my work'. She'll [the teacher] just say, 'Yes I am.' (S53/I/95b)

Only five students resisted this action - they lifted their desk lids and covered their books with their arm.

Eight students contrasted their own inability to manage the disclosure of their ideas with their teachers' ability to choose not to 'bother' to put in sufficient time and effort to find out about student ideas. This was viewed with concern as they considered they benefited from teachers knowing about their learning. Two students asserted:

> S94   Teachers could [find out if students understand] but they don't bother.

> S96   I think they're got to look at their priorities. (SG92/L10/96)

Moreover, they claimed teachers' decisions about whether to 'bother' were influenced by the teacher's perception of a student's attitude. In a way that suggested this view was long held, one student quoted her sister's advice to her:

> ... that's what my sister said to me. 'Don't get into the teachers' bad books .... if they think that you don't want to learn they won't bother with you. (SG92/L5/96)

These students considered teachers had opportunities to help them but that they were able to choose not to do so.

Disclosure and confidentiality and potential harm

Another aspect of teachers' rights in the classroom was their ability to disclose student ideas to others without their consent, and thereby expose the student to potential harm. The students considered this happened when teachers replied to a private question in a public way. The significance of this action was pointed out to the researcher by a student after she had approached a teacher for help and the teacher had replied to the class (SG91/L5/96). The twenty eight students who were asked, described this action as potentially harmful because of the possible responses of their peers. They asserted the *possibility* of this action caused them to restrict the disclosure of their ideas to teachers. In a manner that suggested this strategy had no benefits, some students claimed they felt 'shame' even when the teacher's comments to the class were 'good'.

The teacher action of replying to individual questions to the class highlighted the differences in teacher and student perspectives and experiences. The teachers considered this action to be an efficient strategy for providing timely and effective feedback to all students, including those who may not be prepared to ask for help (Cowie and Bell, 1995). In contrast, the students saw it as a breach of confidentiality, leading to potential harm to their self esteem and relationships. They recommended that all feedback on their learning be kept confidential. A representative comment was:

> Well, if they've got a complaint, they should talk to you quietly. Cause I hate it when they talk out loud and everybody laughs ...then I get smart comments later. (SG81/L6/96)

The students indicated that the possibility of the teacher disclosing their ideas to others, influenced their willingness to ask questions, thereby disclosing their interests and ideas.

Disclosure and expectations of peer and teacher actions

Student comments indicated that they often sought to limit the disclosure of their ideas because of their prior experiences of teacher and peer reactions. Teacher and peer reactions were viewed as unpredictable, in that they were said to act in ways that undermined student self esteem, relationships with others and helping students understand ideas. Expectations of harmful teacher actions appeared to be of long standing as the students drew on the experiences of their parents and siblings to make their point. For example, one student described how her mother's teacher used to embarrass her mother by asking her, in front to the class, to confirm she understood an idea (S54/I/95b).

The students' perceptions that interaction with teachers involved risks was highlighted when Teacher 5 was explaining the comments she had written to the class. She told the students she had asked some of them to, 'See me' (T5/FN3/95b). The class told the teacher they 'hated' this sort of comment because they automatically assumed they had 'done something wrong' and would 'be in trouble' or be 'yelled at' (T5/FN3/95b). One student burst out, 'Teachers are like sharks'. The teacher asked the students if they could recall her or any other teacher within the past two years shouting at them. Only two said they could but they all assured her, 'You never know what will happen.' (T5/FN3/95b). The teacher commented to the researcher that it was a student myth that teachers shouted at students, albeit a powerful one.

Other students asserted that one of the difficulties with interaction was that teachers did not always respond to the content of their questions, or to them, as if they seeking help to understand ideas. They claimed teachers showed displeasure at being asked, implied they were asking because they had not been listening and / or were they were slow to understand. They reported teachers reacted to questions by growling, yelling, shouting, by being grumpy or becoming angry. They described these reactions embarrassing and 'belittling'. One student explained their view clearly:

> The worst thing is when you ask a questions and they [the teacher] belittles you in front of everyone and goes 'Weren't you listening?' or 'Don't you understand that by now?' (SG91/L9/96)

Another student explained:

> Sometimes you ask them and they spend so much time growling you that they never actually explain so you still left wondering 'What are we doing?'. (SG91/L5/96)

Two others said:

> S Teachers sometimes look at you as though your are stupid.
>
> A You deaf or something? (SG71/L8/96)

The students indicated the possibility teachers would 'bite your head off' made them, in the words of one student, 'scared to ask them again' (SG83/L7/96). That is, it made them limit the disclosure of their ideas. Two students said that, given the chance, one change they would make to teachers would be to make them 'easy to approach ... not get mad ... if you don't understand it then not shout at you ... not get frustrated and annoyed you have to ask them again' (S56,57/I/95b).

A related influence on disclosure was the students' perception that quick understanding was valued. This inhibited them from asking questions because of the possibly they would be judged as 'slow' and so restricted their teachers' and their peers' knowledge of what they did and did not understand. Thirty students (all those interviewed from three classes in phase 2) considered teachers expected them to understand ideas and to complete tasks within a specific time. They were aware that teachers valuing of quick understanding may derive from teachers' obligation to teach what was in the curriculum but they claimed not understanding within the prescribed time produced negative feedback from teachers. One student explained:

> It's kind of, they [teachers] set the work and ... if you can do it at the right pace they're doing it, you're OK, but if you can't, you kind of head back and then you get in trouble. (S54/I/95b)

It seemed teacher actions, which valued questions to extend ideas, also led the students to conclude quick understanding was valued and questions that sought further clarification were not. One student described teacher actions thus:

> ... if you ask like an extended question, like thinking ahead to try and add something more difficult into an experiment. The teachers say 'Oh yes, that's a good question'. If you ask something that you didn't understand from before, then that is not a good question. (SG92/L6/96)

The students were concerned therefore that their questions would disclose they were 'last one' to understand and lead to their being judged as 'slow'.

Interestingly, teachers taking time to ensure students understood ideas was characteristic of three occasions which the students reported as being particularly helpful (SG91/L5/96; SG71/L10/96). When asked, what feedback teachers should provide, twenty of the thirty students recommended teachers provide feedback to support effort and persistence. Such feedback has the possibility of countering the myth that learning / understanding should happen quickly and easily.

Disclosure as leading to help

In contradiction with the fears evident in the previous quotes, the students stated teachers and peers could act in ways that helped them understand ideas and so interaction with them was sought and valued. The students expressed the desire for more opportunities to ask questions of teachers and considered it part of a teacher's role to answer their questions. They considered it would be like being away if a teacher did not asses their ideas. In addition, twenty students recommended the main way teachers could enhance their learning was for them to discuss ideas.

Their comments construed the decision involved in asking a question, as dilemma driven. This dilemma was explained by a student after a lesson on the difference between mass and weight. She said:

> That is what it as like today. I kept on thinking that I would put up my hand [and ask a question] but then someone else would put up their hand and they would understand it perfectly and I thought 'Well, everyone else probably understands it and I don't'. ... then I'd look stupid if I put up my hand and asked her to repeat it. She could have already gone over it ten times since I didn't understand it. I'd look like a X for making her explain it once again because everyone understood it. (SG92/L9/96)

In this instance, as the students' friend pointed out, the students asking the questions did not understand the idea. However, at the time, the student the student had not appreciated this and she had prioritised her academic status and relationships with others, over her desire the understand the idea being taught.

Disclosure as mediated by trust

Trust was described by the students as a key interpersonal factor that mediated their willingness to disclose their uncertainties. They preferred to seek help from trusted peers and teachers. The trustworthiness of peers was considered crucial. A representative comment was:

> You need to be able to trust others, to be sure their reactions won't be to make fun, talk about or think I am stupid. (S56/I/95b)

Considerate students and friends minimised the threat of asking for help in the social setting of the classroom, because they could be trusted to be well intentioned. This was illustrated by as student who contrasted the research class with an option class. She explained:

> We know them pretty well but in some option classes you feel like you can't really ask questions because there are other kids who think they are real neat. They do put you down. They look at you and go, 'Why did you ask that?'. You sort of feel uncertain. (SG71/L9/96)

A number of girls elaborated on the extra support they gained from working with friends. One group of four Year 10 girls claimed, and the researcher observed, that they did not question the teacher as individuals. Instead they discussed their problems as a group and one of them asked the teacher for help when an issue was unresolved. This was illustrated by the girl who told the researcher:

> We don't normally like putting up our hand and saying 'I got this answer', we normally say 'Our group'. ... Because we do all our work, basically, together. ... if we put up our hands and say, 'We got this answer' and she realises it is wrong she will come down and talk to us as a group, not individually. (SG71/L9/96)

However, students asking their peers, rather than the teacher, does reduce the teacher's access to student thinking.

The trustworthiness of a teacher's reactions was described as influencing student willingness to interact with teachers in ways that disclosed their thinking. The students explained they formed impressions of teachers' likely actions and reactions 'over a period of time and from what you hear from people'. Three students, interviewed at the beginning of the school year, actively assessed how their new teacher interacted with students. They explained:

S95   In a way I kind of assessed X [the teacher]. It was ... the first lesson where we actually did something and it was kind of like.

S94   Yes.

S95   And it was interesting to see how she was going to go about it and talk to us.

...

S94   I was just kind of sussing out ... how far you could go with

S96   And what her limit was.

S94   Yeah.

S95   If she was prepared to explain it again to you and not just say it once. That's it.

S94   And to treat the class all ... the same ... not certain people.

S96   Get certain treatment or.

S94   Or this one is really bright so she gets special attention and this one is quite
             dumb so.

S96   'I won't waste my time with her', sort of.

S95   But that didn't happen. (SG91/L3/96)

The students asserted they needed to feel 'safe' or 'comfortable' with a teacher
before they asked questions. One student explained how uncertainty about how a
teacher might react, led her to ask her parents:

> ... if you've been with a teacher for awhile, you sort of, you know their reactions and
> stuff, if you feel comfortable asking them. But, like, if you're not really sure, like, this
> teacher I had in primary school ... he was sort of in and out and, you didn't really know
> if you were going to ask him at the wrong time or not ... so I sort of left it till home.
> (S55/I/95b).

It seemed the expectations, which students developed of their particular teacher,
reduced the unpredictability of teacher-students interactions and if the teacher was
trustworthy, it could enhance student willingness to seek help, thereby disclosing
their ideas.

In summary, student disclosure was a crucial aspect of formative assessment.
Without it, formative assessment simply cannot occur. Formative assessment is
determined by the quality and quantity of student disclosure. Section 4.3 has described
those aspects of formative assessment, which determine the degree to which students
will take a risk and disclose what they know and don't know, or can and cannot do.
These aspects are the students' perceptions of the teachers' assessment strategies, the
relationship between teachers' rights and disclosure, disclosure as a source of potential
harm and trust as mediating the disclosure. The important and complex task for the
teacher is therefore to mediate the social context of the classroom to ensure that the
risk to students of disclosing is minimised. For this, the teacher's ability to monitor
and use the power and authority relationships in the classroom is crucial.

4.4   A TACIT PROCESS

A fourth characteristic of formative assessment was that it was often seen as a tacit
process. A frequent comment from the teachers was that they were not always
consciously aware of doing formative assessment, and in particular unplanned or
interactive formative assessment, for example:

> I am still not recognising what I do in terms of unplanned (interactive formative
> assessment) (TD5/96/14.1)

The teachers tacitly undertook formative assessment and were not always able to
explicitly describe it to the researchers. This unawareness was evident in the
discussion by the teachers on the use of 'gut-feelings', for example:

Because you can stay awhile with a group ... oh, I'll just listen to the kids. And that's where you get your gut assessment. (TD5/96/15.24; TD5/96/15.25)

And you don't get a gut feeling sitting the night before thinking about ... It's when you're there, so you are interpreting something that's happening in the room ... That's what your gut feeling is. ...How many times do you actually change your teaching style or whatever, during the lesson, because the gut feeling gives ..... You've said something and you know exactly where you want to go with the kids, and something sort of happens, and it's not working, so you sort of get that gut feeling. You think pretty fast then. ... But you can't tell other people that you work by gut feelings, because they need something tangible that they can actually ... think about and something you can rationalise. You can't rationalise just in cold turkey gut feelings. (TD7/96/20.39)

The teachers also spoke of 'getting an impression' of the class, for example:

And the formative assessments .. could be .. more formal tasks or they could be just impressions in the classroom .. we cant really identify what tells us that the majority of the class know the first bit so we can go onto the second bit, but to me, that automatic assessment is part of formative assessment (TD4/95/11.30; TD4/95/11.31)

The experiences of being involved in the research had made more visible to the teachers what formative assessment information they were collecting and what they were doing with it, for example:

I personally never really realised I was doing it (formative assessment) except that the class was with me, or not with me. And since this experience, you sort of tend to focus more on what am I actually taking in here. Or what is it actually telling me .. this process is going on. And I think for most teachers it will still be subtle and not obvious.(TD9/96/27.3)

The teachers stated that thinking about formative assessment had helped them become more aware of their professional knowledge and skills and more able to use these in the formative assessment process in the classroom.

4.5 USING PROFESSIONAL KNOWLEDGE AND EXPERIENCES

A fifth characteristic of formative assessment was that it was seen to rely on a teacher's professional knowledge and experiences. The professional knowledge and experiences of the teachers were seen as important in attending to some sources of information (rather than others), in interpreting the elicited information and in taking action. This professional knowledge and experience included the teachers' knowledge and experiences of the topic, of the students as learners, and from having taught the unit of work before. For example:

Because the knowledge of how to teach is what makes that successful. ... you've got to have confidence in your ability to teach.

And that's all your other skills ... those you actually can't do without.

...Yeah, it's knowing how to handle the students, isn't it?

I still think that the knowledge base of the subject has got a place, though. I mean,

... I think, too, if your knowledge is at a reasonable level, you can take advantage of the one off situations that sometimes happen. Whereas if it's not there, you can't take advantage at all.

And the more you teach, the better you become. (TD9/96/26.41)

4.6 AN INTEGRAL PART OF TEACHING AND LEARNING

Another characteristic of formative assessment was the action taken by the teacher and the student as a result of the information gathered. Taking action to enhance learning is an integral part of the definition of formative assessment. The teachers commented on the variety of actions they took in response to the formative assessment information as well as the way in which they evaluated their actions. For example:

> I was thinking the teacher would get some information, interpret it, decide to act, they would act, and then it may or may not work, and they would react to that. ... Deciding on a new experiment, deciding to do a discussion, what sort of teaching reaction. ... To make a decision where you go. Are you going to reteach it or are you going to have a look at it from a different (perspective), go get some more information. ... Move on to the next step, cause they've got it. ... you interpret information see, then you act on, then you react.(TD7/96/20.44)

Taking the action involved the teachers making decisions and judgements, using their professional knowledge and experiences. The action often appeared to the teachers to be a part of teaching and the comment was made as to whether the action was a part of teaching or a part of assessment. The overlap between the action inherent in formative assessment and teaching was frequently acknowledged, for example:

> I think formative assessment and teaching .. overlap really (TD5/96/14.9)

The teachers described their actions as those to facilitate students learning. They spoke of actions that mediated the students learning of the science and actions which enhanced the personal and social development of the students. The actions taken were, for example, suggesting further questions, suggesting further activities, questioning of a student's ideas, explaining the science, giving feedback as to the students scientifically acceptable or unacceptable ideas. The notion of the teacher as a neutral facilitator was not seen as part of formative assessment:

> Being a neutral facilitator isn't what we're on about here. In terms of formative assessment you (are) wanting to take action, you may choose to do nothing because you want to leave the kids for a while to see if they can find their way through it, but if they can't, you might want to then make another decision. (TD7/96/20.67)

A similar comment could be made as to whether the action taken by the students was a part of learning or assessment. These comments highlighted that formative assessment is an integral part of teaching and learning.

4.7 WHO IS DOING THE FORMATIVE ASSESSMENT?

Another characteristic of formative assessment was both the teacher and the student were doing the assessing. The teachers' comments highlighted the involvement of students as assessors, in addition to the teachers. The following is part of a discussion of the model of formative assessment, and in particular the cycle of gathering information, interpreting, and taking action:

> Think of it from the kid's point of view, the kid gathers information from what you've given them already, they filter it, decide what's relevant to them, they interpret what they need to do however they like, they act on that information, and then from whatever you do or from whatever things happen, they gather more information and so on.

> So it works exactly for them. It's just that our acting becomes their gathering information points.

> Whatever we do .... they get the information from (it).

> And their acting is our gathering.(TD7/96/20.45; TD7/96/20.59)

Student self-assessment was seen as an important part of formative assessment:

> Formative assessment isn't just for the teacher, it's for the students to know that they are still moving, and going somewhere ... So its a decision making process for the student (TD4/95/11.38)

Some teachers used the phrase 'self-assessment' to refer to the teachers evaluating their own teaching:

> I think it's self-assessment by the teacher as they go along. As well as the other side which is helping the students assess themselves ... I think it's what we automatically do - assessing ourselves as we go along .. and the kids ... assess themselves (TD4/95/11.40)

4.8 THE PURPOSES FOR FORMATIVE ASSESSMENT.

The eighth characteristic of formative assessment was the purposes for which it was done. As documented in the case studies, the teacher development days and the surveys (see Bell and Cowie, 1997), the teachers identified several purposes for formative assessment. In particular, they identified that the two main purposes of formative assessment were to inform the students' learning and to inform their teaching.

The purposes to support the students' learning included monitoring the progress, learning or understandings of the students, during the teaching and learning, for example:

> .. The teachers' purpose for the thing that they're doing at the moment, what they want the children to learn, what they are trying to get out of them, the kind of thing that they do (TD7/96/20.31)

The learning might be social, personal or science learning (Bell and Cowie, 1997):

> So in that purposes for learning, ... you do have science purposes ... You did have a
> social purpose and a personal purpose (TD7/96/20.55)

The purposes to support learning also included giving feedback to students about what learning was valued in the classroom, giving legitimacy to the students scientifically acceptable ideas, supporting long or short term goals and finding out whether an activity or task was 'working' or not, for example:

> ... Is there a case when you just check to see whether the activity is working or not?
> Like, I mean, just thinking from my own teaching where you might set up a group
> activity and you realise after visiting two or three groups that the instructions haven't
> been clear enough, so you stop the class and say, look instruction number 3, I've
> actually missed a bit out. It should read like this, and everyone nods and away you go.
> So in this case you haven't actually checked that, you haven't gone right back to the
> learning goals, you've just got to the level of 'is this activity working or not'.

> I was just thinking on that train of thought too because it's sort of as if we're a trouble
> shooter and just watching if things are moving in the right direction ... You have to
> intervene in some way. It could be to the whole group or an individual.

> But how do you do it. I suppose, like we said, do you stop the group or do you speak to
> the individual on the side..... (TD7/96/20.71)

The purposes to support teaching (mentioned by the teachers) included planning in the current lesson and unit; planning for future teaching; knowing when to input new ideas and when to move on to the next topic; knowing when to introduce an activity to maintain interest and motivation; evaluating the actions taken in previous formative assessments and teaching activities; finding out if the students had understood or not; providing information to report to students, caregivers and the school; and providing assessment information additional to the quantitative marks on achievement in reporting.

4.9 THE CONTEXTUALISED NATURE OF FORMATIVE ASSESSMENT.

The ninth characteristic was that formative assessment undertaken by the teachers and students was contextualised. In other words, the purposes, the information elicited, the interpretations made, the actions taken depended on many contextual factors. For example, the ways the formative assessment information was elicited, interpreted and acted on was influenced by the learning situations used (whole class, small groups or individuals); by the learning activities chosen (for example, brainstorms, investigations, watching a video, library projects); the teacher's knowledge of the students; the professional knowledge and skills of the teacher; the topic of the lesson and the teacher's purposes for the lesson.

4.10 DILEMMAS

The tenth characteristic of formative assessment was that of the dilemmas faced by the teachers when doing formative assessment. The interaction between these characteristics in the processes of formative assessment presented the teachers with dilemmas. The word 'dilemmas' is used as there was no obvious solution to the situation and the decision made in response to each situation would depend on contextual features and the teacher and students concerned. Unlike problems which can be solved, dilemmas are managed and this management relies heavily on the professional judgement of teachers. The nature of these dilemmas was evident in the discussions on the teacher development days on the tensions between formatively assessing the class or an individual; between formatively assessing the science or the personal and social development; between formatively assessing the science in the curriculum and the science outside the curriculum; and between the different purposes for eliciting and taking action.

4.11 SUMMARY

In summary, the ten characteristics of formative assessment that were identified by the teachers and students were that formative assessment is seen as being responsive; it is often a tacit process; it relies on student disclosure; it uses professional knowledge and experiences; it is an integral part of teaching and learning; it is done by teachers and students; it is a highly contextualised process; and it involves the management of dilemmas. Important considerations are the sources of evidence, including student disclosure and the purposes for which formative assessment is done. All-in-all, formative assessment is a highly complex and skilled activity for both the teacher and the student. Formative assessment is not something teachers are likely to learn to do in a short session in an inservice course. It is a professional skill that develops with increasing professional experience, awareness and reflection.

Another way to summarise the data from the case studies and the teacher development day discussions, was to model the process of formative assessment. One such model is discussed in the next chapter.

CHAPTER 5

A MODEL OF FORMATIVE ASSESSMENT

In the previous chapter, the data were summarised as ten characteristics of formative assessment. In this chapter, the findings are summarised in the form of a model to describe and explain the formative assessment processes as carried out by the teachers (Cowie and Bell, 1999). The model presented here was developed from a consideration of the data collected during the case studies, including the interviews, classroom observations, surveys; the current literature; and the debates on the teacher development days. In particular, the development of the model has drawn heavily on the discussions by the teachers on the teacher development days, when they addressed the task of developing a model for formative assessment (Bell and Cowie, 1997, p. 277) to describe what they were doing, to other teachers.

5.1   WHAT WAS ASSESSED IN SCIENCE CLASSROOMS?

One consideration in developing a model for formative assessment was what was assessed in the science classrooms observed. What was being assessed was related to the purposes of formative assessment. The students' learning within the classroom involved their personal, social and science development (Cowie, Boulter, Bell, 1996, p.30). Students' personal development related to their learning about themselves as learners of science, asking questions, self-assessment, behaviour, time management, motivation and attitude. Their social development related to their interacting with others (students and teachers), peer assessment, leadership skills, group work, discussion and listening skills. Their science development related to the development in the knowledge and understanding of science and their ability to do science - for this was their unique purpose for being in a science classroom. These three aspects were assessed by the teachers of science (Cowie and Bell, 1996). The three aspects were not independent of each other, the complexity and richness of their interactions was a contributor to the diversity and complexity of the classroom. The three aspects are conceptualised as three intersecting circles on the left-hand side of figure 1:

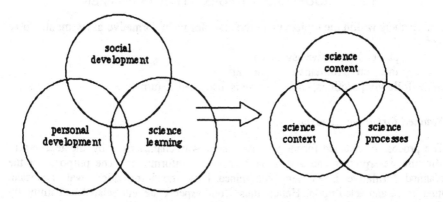

Figure 1: What is assessed in science classrooms?

The aspects of science which were assessed in the science lesson are represented on the right hand side of figure 1, namely science content (the body of scientific knowledge - the concepts and ideas of science), science context (the contexts in which the science is learnt and used) and science processes (those skills and processes used by scientists to investigate phenomena).

The balance of assessment of personal, social and science development or learning was discussed by the teachers (TD3/95/7.10). [A full explanation of the data codes is given in the appendix]. It was acknowledged that the three aspects may differ in weighting within a lesson and within a unit of work (TD3/95/7.11; TD3/95/7.17; TD3/95/7.18); with different learning goals for a lesson (TD3/95/7.11) and with different abilities and ages of the students (TD3/95/7.9). The analogy was made between the circles and balloons that inflate and deflate (TD7/96/20.15).

The teachers commented that the way they assessed the social, personal and science aspects was different. The secondary teachers felt that science assessment was done more by formal methods and the assessment of the social and personal by more informal means (TD3/95/7.24), for example:

> I would tend to assess ... the three science areas quite formally over the year. You know, there were things that you would look at specifically over time. The personal and social skills, it's only an informal assessment ... apart from comments that I would make to parents or whatever, or to the kids, I don't need to actually have some ... formal assessment of those things. But I will be aware of how somebody is developing, how that they are now discussing things or whatever, how they are changing their attitudes to the way they are learning. So it's an informal kind of thing, rather than specifically structuring some kind of assessment to do it ... To me it's getting back to the gut feeling and just noticing things....(TD3/95/7.19)

However, the primary teachers indicated that they assessed by formal ways both the personal and social learning (TD3/95/7.22, TD3/95/7.25) as well as the science learning.

5.2   MODELLING OF FORMATIVE ASSESSMENT.

The teachers within the project undertook two forms of formative assessment. These were:
- planned formative assessment
- interactive formative assessment

In the following sections, each of these is discussed in turn.

Planned formative assessment

The process of planned formative assessment was characterised by the teachers as eliciting, interpreting and acting on assessment information. The purpose for the planned formative assessment determined how the information was collected, interpreted and acted upon. Hence, these four aspects are interrelated and mutually determining. These aspects can be represented diagrammatically as:

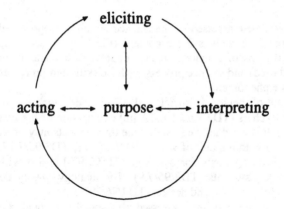

Figure 2:   *Planned formative assessment*

The planned formative assessment process was seen by the researchers and teachers as cyclical or spiral. For example:

> We decided it was a cycle, and the cycle starts with a student activity of some sort, data gathering happens ... the teacher needs to be reflecting on what is happening.. the teacher and the students will formulate the direction that they're going, and then you're back to the beginning with a student activity of some sort... there is feedback going on ... (TD4/95/12.6)

Each of the four parts of the process are described in more detail.

Purpose

The main purpose for which the teachers said they used planned formative assessment was to obtain information from the whole class about progress in learning the science as specified in the curriculum. This assessment was planned in that the teacher had planned to undertake a specific activity (for example, a survey or brainstorming) to obtain assessment information on which some action would be taken. The teachers considered the information collected as a part of the planned formative assessment was 'general', 'blunt' and concerned their 'big' purposes. It gave them information which was valuable in informing their interactions with the class as a whole with respect to 'getting through the curriculum'. This form of formative assessment was planned by the teacher mainly to obtain feedback to inform her or his teaching. The purpose for doing the assessment strongly influenced the other three aspects of the planned formative assessment process.

Eliciting information as a part of planned formative assessment
The teachers planned in advance to elicit information on their students' science understandings and skills learning, using specific assessment tasks. While the teachers described their eliciting of planned formative assessment information as purpose driven, their purposes for eliciting information often changed during the unit of work.

At the beginning of a unit of work, the teachers frequently planned to assess their students formatively to inform their planning and teaching during the rest of the unit. For example, teacher 7 asked her students to write down, in a brainstorm activity, all they knew about how hot-air balloons worked so that she could find out what the students already knew about the area of science she was intending to teach.

During a unit of work, the teachers planned to elicit formative assessment information, using specific assessment activities, on the understandings their students were constructing during the teaching and learning. The formative assessment tended to focus on the extent to which the students had learnt what the teachers were intending the students to learn. They often did this at the beginning of a lesson, and used the information during the lesson. For example, teacher 3 used quick tests at the beginning of each lesson during a unit of work to find out what the students had learnt and remembered from the previous lesson.

The teachers also planned to elicit formative assessment information at the end of a unit. They used this information in a formative way to inform their teaching when they taught the unit again. For example, teacher 7 had taught the hot-air balloon unit before and was able to anticipate some of the students alternative conceptions in her initial planning.

The teachers said the strategy they used to gather the information was determined by the nature of the information they required (and therefore indirectly the purpose). For example, the teachers used strategies such as asking quick questions to obtain information on students' on-going understanding and recall; brainstorms to obtain information on the scope and depth of the students' prior knowledge; asking the students to generate questions to obtain an indication of what they were interested in finding out; making physical models to elicit information which did not depend on the students' knowledge of scientific vocabulary; and asking students to record their explanation of how an event occurred to gain information on students' conceptions.

Within planned formative assessments, the teachers usually planned to use assessment strategies to elicit permanent evidence of the thinking of each individual student. For example, the teachers usually asked the students to write something on paper or to make a physical model. This meant these assessment occasions tended to be semi-formal.

Interpreting as a part of planned formative assessment

The second aspect of the teachers' planned formative assessments was that of interpreting the information. The purpose for the planned formative assessment was influential on the process of interpreting. As the teachers' planned assessments tended to have a science curriculum focus, their interpretations were usually science criterion-referenced. That is, the teachers wanted to know if the students had learnt and understood the science they intended them to learn. For example, through spot tests, teacher 3 elicited formative assessment information which he interpreted as indicating some of his students not having understood certain ideas on electricity from the previous lesson.

The teachers' interpretations of their planned formative assessments also had an element of norm-referencing in that they were influenced by their expectations of the understanding which was likely with students at a particular age or year of schooling. For example, when teacher 5 interpreted her students' understanding about the idea of tectonic plates on the earth's surface, she took the concept no further given they were 12 year olds.

The teachers indicated that their knowledge bases (Shulman, 1987) were important factors in their being able to interpret the information they had collected in their planned formative assessment. They indicated that interpreting involved using their content knowledge; general and content pedagogical knowledge; curriculum knowledge of learners and their students in particular; knowledge of educational contexts and a knowledge of educational aims and goals. The teachers indicated that their experience of teaching a particular concept added to their pedagogical knowledge and hence their ability to interpret student thinking when teaching the concept on a further occasion.

As part of the interpreting, the teachers felt that they filtered out the irrelevant information - information that was seen to be irrelevant to the purpose of that particular formative assessment episode, for example:

> The gathering of information ... it's a sifting activity, isn't it. To see if it's relevant to what your goal. (TD5/96/15.32)

The ability to discriminate relevant from irrelevant and to recognise the significance of something was important. Prior teaching experiences were considered by the teachers to be important in the ability to discriminate in the process of formative assessment. In other words, the teachers considered formative assessment to be more likely to be undertaken by more experienced teachers, and later in the teaching year, rather than earlier.

Acting as a part of planned formative assessment

Once the teachers had elicited and interpreted the information they had available, they had the opportunity of taking action to enhance the students' learning. Action on the interpreted information is the essential aspect of formative assessment that distinguishes it from continuous summative assessment. To do this, the teacher needed to plan to have a flexible programme and to allow for ways in which she or he could act in response to the information gathered. It also helped to be able to act in a variety of ways in response to that gathered information. Hence, the teacher's pedagogical knowledge informed the taking of action as part of the process of formative assessment.

The teachers acted on the assessment information they obtained in three different ways: science-referenced, student-referenced and care-referenced. That is, the purpose for their taking action varied and was often broader than their purposes for eliciting the information, which were mainly science based.

(i) science-referenced ways of acting

One way the teachers acted on the planned formative assessment information was to mediate the students understandings towards those of scientists by addressing students' alternative conceptions. For example, teacher 7 decided on the five quick practical activities she would use once she had found out the students' alternative conceptions.

In another form of the science-referenced way of taking action, the teachers acted on the planned assessment information to ensure that the students performed the task in the manner they had prescribed. The teachers considered which of the tasks they had selected would mediate the students' understanding towards a particular science concept. They also wanted to ensure the students had common experiences. For example, teacher 7's five short practical activities provided the students with opportunities to gain common experiences which could then be discussed in class.

A third way the teachers acted on planned assessment information in a science-referenced way was to indicate what was valued as a learning outcome. For example, teacher 7 legitimated the scientist's view of density by recording it on the board after a class discussion.

(ii) student-referenced ways of acting

The second way of acting on the planned formative assessment information was to act in a way that was referenced to the individual student. In particular, the teachers used student-referenced action that built on how an individual student's science understandings were changing over time. For example, teacher 8 provided students who had finished a ray box activity, with an extension activity of using two prisms 'to produce a rainbow and then combine the colours'. The action was given with reference to these students' need for further focused activity and learning.

(iii) care-referenced ways of acting

The third way the teachers acted on the planned formative assessment information was to act in a way to sustain and enhance the quality of interactions and relationships between the students and between themselves and the students.

In summary, the teachers used planned formative assessments and in doing so planned and undertook the eliciting of information. They then interpreted and acted on that information. In this section, each of these four aspects of planned formative assessment was discussed as a separate entity. However, in looking at the interaction between these four aspects, three further points about planned formative assessment can be made:

• The time taken for the process varied.

• The purposes for eliciting planned formative assessment information could be different from those for taking action on that elicited information.

• Both the students and the teachers contributed to planned formative assessment and in a reciprocal way- when the teacher was taking action, the students were eliciting information and when the students were taking action, the teacher was able to elicit information.

Interactive Formative Assessment

The second form of formative assessment was interactive formative assessment. Interactive formative assessment was that which took place during student-teacher interactions. It differed from the first form - planned formative assessment - in that a specific assessment activity was not planned. The interactive assessment arose out of a learning activity. Hence, the details of this kind of formative assessment were not planned, and could not be anticipated. Although the teachers often planned or prepared to do interactive formative assessment, they could not plan for or predict what exactly they and the students would be doing, or when it would occur. As interactive formative assessment occurred during student-teacher interaction, it had the potential to occur any time students and teachers interacted. The teachers and students within the project interacted in whole class, small group and one-to-one situations.

The process of interactive formative assessment involved the teachers noticing, recognising and responding to student thinking during these interactions and can be represented diagrammatically as:

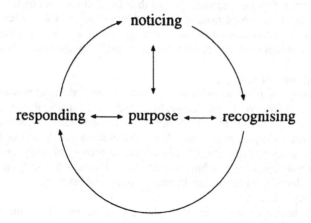

Figure 3: Interactive formative assessment

Purpose

The main purpose for which the teachers said they did interactive formative assessment was to mediate in the learning of individual students with respect to science, social and personal learning. Hence, they said they formatively assessed a wider range of learning outcomes than the science and this is in line with the *New Zealand Curriculum Framework* (Ministry of Education, 1993a). The teachers' specific purposes for interactive formative assessment emerged in response to what sense they found the students were making. The purposes for the interactive formative assessment were an important part of noticing, recognising and responding. Interactive formative assessment was therefore embedded in and strongly linked to learning and teaching activities.

The teachers indicated that through their interactive formative assessment, they refined their short term goals for the students' learning within the framework of their long-term goals. For example, teacher 5 changed her purposes for learning, within a unit of work on earth sciences, from learning about weathering to learning about contracting and expanding when she noticed some of the students had scientifically unacceptable conceptions about heating and cooling.

The teachers indicated that their purposes for learning could be delayed. For example, teacher 7 delayed the learning about separating mixtures until the students had learnt about the properties of the substances to be separated. The purpose of student learning was negotiated between the teacher and the students through formative assessment feedback. Teacher 9 described it as linking students into her agenda. The teachers described interactive formative assessment as teacher and student driven rather than curriculum driven. They said the focus of their interactive formative assessment, was 'finer tuned' with 'lots of little purposes to support the major picture or purpose'.

Noticing as a part of interactive formative assessment

Noticing was a key part of the interactive formative assessment. Noticing in interactive formative assessment differed from eliciting information in planned formative assessment in that the information gained was ephemeral, not recorded, and the noticing was faster than the eliciting. The ephemeral information the teachers gained was verbal (student comments and questions) and non-verbal (how they did practical activities, how they interacted with others, the tone of discussions, their body language). This information was of thinking and actions in progress. If the teachers were not present when the information was noticeable, the information was rarely available to the teacher at some later time. The teachers were able to access some permanent evidence of the students' thinking by reading their written work. Some student comments, questions and actions were relatively transparent in that what the students said or did implied a particular way of thinking. For example, teacher 3 asked his students to connect a light bulb, wires and a dry cell to make the light bulb glow. By watching the students, he could see how the students used the wires and thereby gain some insight into how the students viewed electric current. However, it must be acknowledged that not all thinking of the students was accessible to the teacher. Through their interactions with the students, the teachers were able to notice information on some but not all of the students. The teachers noticed different information from different students at different times.

The information they noticed was related to the students' science, social and personal development (Bell and Cowie, 1997). The teachers noticed information about the students' science learning. During practical and written work, the teachers noticed whether or not the students were performing the task according to the scientific procedure, and if they were making the intended sense. For example, teacher 5 noticed that some of her students did not have a scientifically acceptable concept of heating and cooling; teacher 9 noticed that some students did not understand the notion of controls; and teacher 7 noticed that some of her students did not know about the properties of the substances she had asked them to separate. The teachers also noticed information on what sense the students were making (whether it fell within their intended learning or not). For example, teacher 7 noticed that some of her students requested equipment to filter oil and water to separate them.

Their request gave her information about the understandings these students were developing from an activity on separating substances. The teachers also noticed aspects relating to the students' personal and social development. For example, teacher 9 noticed when a student began to work more co-operatively with others.

Recognising as a part of interactive formative assessment

The second part of interactive formative assessment was that of recognising. The teachers commented that while they were observing, talking to or listening to a student(s) they would notice something and recognise its significance for the development of the students' personal, social or science understandings. Recognising may be differentiated from noticing in that it is possible to observe and note what a student does without appreciating its significance. At times, the teachers only appreciated the significance after the event. For example, teacher 3 was confused by how one of his students used the word 'fuse', seemingly with understanding. It occurred to the teacher later that the student was using the term 'fuse' in the context of a bomb fuse, rather than to the teacher's meaning of the fuse in a household circuit.

The teachers also said their noticing and recognising was strongly influenced by their pedagogical knowledge and experiences from previous teaching. Noticing and recognising required the teachers to use their prior knowledge of the individual student, their pedagogical content knowledge, and their knowledge of the context.

The teachers discussed how they might not notice and recognise the significance of students' comments and actions because they lacked the experience and knowledge. The teachers mentioned that interactive formative assessment was difficult for beginning teachers and for experienced teachers with a new class, say at the beginning of the year (Bell and Cowie, 1997). The teachers said their awareness of student thinking (what they noticed and recognised) was often triggered if a student response was unexpected, incorrect or a number of students indicating that they held a similar view.

Wiliam (1992) identified two issues, disclosure and fidelity, which may limit the information teachers have to notice and recognise. The disclosure of an assessment information eliciting strategy is the extent to which it produces evidence of attainment from an individual in the area being assessed. Within this project, this definition is extended to include the extent to which it produces evidence of students' non-understanding in the area being assessed. Wiliam (1992) defined 'fidelity' as the extent to which evidence of attainment, which has been disclosed, is observed faithfully. He claimed that fidelity is undermined if evidence of attainment is disclosed but not observed. For example, the teacher may not hear a small group discussion in which the students demonstrate they understand a concept. Fidelity is also undermined if the evidence is observed but incorrectly interpreted, for example, if the teacher did not understand the student's thinking, a possibility if there is insufficient commonalty in the teacher and the students' thinking. The compromise to fidelity through lack of opportunity to observe student thinking and the inability to interpret students thinking, are critical issues in interactive formative assessment.

To recognise the significance of what is noticed, teachers must be able to interpret the information they have and to understand its implications in terms of what sense the students are making. The two parts of interpreting and appreciating implications constitute recognising. From a constructivist view of learning, recognising the significance of what students say and do, requires teachers to make 'qualitative judgements' (Sadler, 1989). Sadler (1989) claimed the use of qualitative judgement required a concept of quality appropriate for the task and the ability to judge work in relation to this. In the case of students' science learning, teachers must understand the science concept (have the science content knowledge) and be able to judge the students' comments and /or actions in relation to this (have the appropriate pedagogical content knowledge). Hence, the teachers must be able to make science criterion-referenced judgements.

Sadler (1989) also argued that qualitative judgements are holistic and invoke fuzzy criteria which are context dependent rather than predetermined. In his view, the salience of particular criteria depends on 'what is deemed to be worth noticing' at a particular time. Wiliam (1992) argued that some forms of understanding can not be completely encapsulated by a set of pre-determined criteria. He claimed that 'construct-referenced' interpretations are required when assessing holistic and open-ended activities such as investigations, projects and creative writing for in these 'The whole is more than the sum of the parts'. Sadler (1989) and Wiliam (1992) both stated that the criteria required to judge or recognise what students are learning as emergent rather than completely pre-determined.

Responding as a part of interactive formative assessment
The third part of interactive formative assessment was the teachers responding to what they had noticed and recognised. The responding by the teachers was similar to the acting in planned formative assessment except that the time frame was different - it was more immediate. Hence, the teachers' responses to student learning were typically care, student and science referenced. A single response might have one or more of these aspects in it.

The care referenced response was a response related to nurturing the teacher's relationship with the students or the students' view of science. The teachers often indicated they would not wish their response to damage their relationship with a student or to damage the way the student viewed science, and at times acted to nurture these relationships. The student referenced response was a response related to enhancing the students' development, with reference to the students themselves. These two were often identifiable in the one response. For example, teacher 7 noted and recognised one student's reluctance to accept her advice on how to separate two substances. He had suggested separating sand and salt with a pair of tweezers. Knowing he would not accept her word on the practicality and preciseness of this proposed method, she gave him a pair of tweezers which he used over two lessons to find out for himself if his technique 'worked' or not.

The science referenced response was a response to mediate a students' learning towards that of a scientist. For example, teacher 5 questioned students and suggested further activities to get them to engage them in thinking about their ideas of heating and cooling.

On some occasions, the teachers responded to interactive formative assessment information by changing from interacting with a random sample of students to acting with all the students. This occurred when a student had a particular scientifically unacceptable conception or when a number of students displayed similar alternative conceptions of the science or of the purposes or requirements of the task. They described this action as being an efficient use of their time and as enabling them to provide feedback to all students including those who might be too shy to ask or unable to formulate their concern into a question. The interactive formative assessment response that they took often involved repeating an explanation or activity, which had been successful with a student or a small group.

Another response by the teachers was deliberately to elicit information from all the students. For example, this was observed during a whole class discussion when teacher 7 responded to the information she had by deliberately eliciting information from all the students by asking for a show of hands. She described the decision to do this as spontaneous. In order to respond to interactive formative assessment in this way, the teachers had to be familiar with a number of assessment strategies and with the forms of information they were able to elicit. When a teacher moved from randomly noticing information from a student in one lesson to eliciting it from all students in the next lesson, this was considered as a move to planned formative assessment.

Responding involves 'reciprocity and empathy' (Learvitt, 1994, p. 73) and spontaneity and flexibility (Goodfellow, 1996). It also involves using prior experiences. Peterson and Clark (1978) found that experience provided teachers with more alternatives for action and Jaworski (1994) suggested that experience can improve teachers' ability to decide which alternative is the most appropriate in a situation.

Within the process of interactive formative assessment, the teachers often had to make quick decisions in circumstances in which they did not have all the necessary information. As these decisions were made in context, the teachers were able to use their knowledge of individual students and the context to help them 'fill in the missing bits of information' (Denscombe, 1995, p 177). Jaworski (1994) claimed that 'teacher wisdom' rather than intuition or instinct is what a teacher brings to this moment of decision making.

In summary, eight points can be made about interactive formative assessment.

• Unlike planned formative assessment, which elicited information mainly on the students' science learning, interactive formative assessment focused on the whole student as it enabled the teacher to focus on all three aspects of students' learning - the students' personal, social and science development (Bell and Cowie, 1997).

• The teachers' pedagogical knowledge bases (Shulman, 1987) were used in all four aspects of interactive formative assessment.

• The teachers indicated that they were prepared to do interactive formative assessment in a lesson. They prepared for it by planning to increase the number of interactions between them and their students. They prepared by providing opportunities for students to approach them. They prepared for interactive formative assessment by rehearsing their responses to possible student alternative conceptions. They also planned to increase their opportunities for observing students interacting with each other.

• Interactive formative assessment depended on the teachers' skills of interaction with the students and the nature of the relationships they had established with the students.

• The teachers viewed interactive formative assessment as an integral part of teaching and learning, not separate from it. The responding as an action could be viewed as a part of formative assessment or a part of teaching from this perspective.

• The degree of awareness of being engaged in the process of interactive formative assessment varied amongst the teachers. And the degree of awareness of the teachers was influential in the processes of noticing, recognising and responding.

• Another aspect of interactive formative assessment was that the process of interactive formative assessment was typically 'over in the moment'.

• The students were not treated equally as they were within planned formative assessment. Typically, only some students were involved in interactive formative assessment at any one time.

5.3   AN OVERVIEW OF THE MODEL OF FORMATIVE ASSESSMENT

In this section, the two forms of formative assessment are discussed together. The two kinds of formative assessment and the links between them, can be represented diagrammatically as:

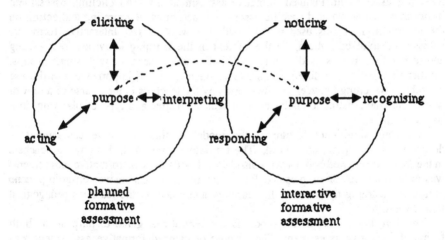

Figure 4: A model of formative assessment

In a meeting in November, 1996, the teachers discussed the relationship between and the interaction of the two kinds of formative assessment (for a fuller account, see Bell and Cowie, 1997, p. 314). They commented that the two kinds of formative assessment were linked through the purposes for formative assessment (see the dotted line in figure 4); that some teachers used interactive formative assessment more than other teachers; and that a teacher moved from planned to interactive and back. The link between the two parts of the model was seen to be centred around the purposes for doing formative assessment.

The teachers confirmed that they changed from planned to interactive formative assessment by noticing something in the course of planned formative assessment. They may have suspected that things may not have been okay and wanted to check things out, they may have noticed a student's or a group of students' alternative conceptions or misconceptions, they may have wished to follow up a hunch, or monitor the learning occurring. This change was usually in response to focusing from the class to an individual. They usually switched back from interactive to planned formative assessment in response to their responsibility for the whole class' learning.

They also commented that under stress (for example, implementing a new curriculum or when ill) they tended to do less of the interactive formative assessment. In particular, heavy emphases on summative assessment procedures for leaving qualifications (for example, Unit Standards in New Zealand) or for review and monitoring procedures (by the Education Review Office in New Zealand) were seen by the teachers as influencing the amount of interactive formative assessment they felt they were able to do.

5.4 DISCUSSION AND SUMMARY

There are five key features of the model. Firstly, the model of formative assessment developed consisted of two kinds of formative assessment: planned and interactive formative assessment. Planned formative assessment involved eliciting assessment information using planned specific assessment activities, interpreting and acting on the information. It was used mainly with the whole class. Interactive formative assessment involved noticing in the context of the learning activities, recognising and responding. It was used mainly with individual students or with small groups. Further similarities and differences are given in table 1. Of importance was the use of and the switching between the two forms by the teachers in the course of a unit of work. The main distinction between them was the degree and type of planning done by the teachers.

A second significant feature of the model is that formative assessment is described as a complex, skilled task. The formative assessment done by the teachers in the research reported here (when considering both forms of formative assessment) was either planned for or prepared for, contextualised, responsive, on-going, done during the learning to improve the learning, and relied on the teacher's pedagogical knowledges.

The third key feature of the model is the central role given to purpose in both forms of formative assessment. The purpose of planned formative assessment was perceived as obtaining information from the whole class about progress in learning the science as specified in the curriculum to inform the teaching. The purpose of interactive formative assessment was perceived as mediating in the learning of individual students with respect to science, personal and social learning. Purpose influenced each of the aspects of planned formative assessment (eliciting, interpreting and acting) and interactive formative assessment (noticing, recognising and responding). The purpose underlying each aspect of one form of formative assessment could differ. For example, the purpose for eliciting and noticing might not be the same as that behind the acting and responding. The purposes were the link between teachers switching between planned and interactive assessment.

A fourth significant feature of the model is the action taken as part of both planned and interactive formative assessment. The action means that formative assessment can be described as an integral part of teaching and learning and that it is responsive to students. The teachers in the research made the claim that they did not think they could promote learning in science unless they were doing formative assessment (Bell and Cowie, 1997).

Table 1: Planned and interactive formative assessment

Planned formative assessment	Interactive formative assessment
the parts of the process were eliciting, interpreting, acting	the parts of the process were noticising, recognising, responding
tended to be done with all the students in the class	tended to be done with some individual students or small groups
could occur over an extended time frame	happened over a short time frame
purposes were mainly science referenced	purposes were science, student and care referenced
responsive to 'getting through the curriculum'	responsive to student learning
what was assessed was mainly science learning	what was assessed was science, personal and social learning
the assessment information obtained was product and process	the assessment information obtained was product and process but ephemeral
interpretations were norm, science and student referenced	recognising was science, norm and student referenced
actions were science, student and care referenced	responses were science, student and care referenced
relied on teachers' professional knowledge	relied on teachers' professional knowledge

The fifth key feature of the model is that the detailed data generated by the research and underlying the model is a valuable contribution to the existing literature on formative assessment. Knowing about the details of the formative assessment process raised the awareness of the ten teachers about what they do by way of formative assessment in their classrooms (Bell and Cowie, 1997). That is, the teachers were doing formative assessment but they were not always aware of exactly what they were doing that could be called 'formative assessment'. The increased awareness enabled the teachers to reflect in new ways on their practice. The increased awareness was perceived by the teachers to be the main aspect of their teacher development during the two years of the research project (Bell and Cowie, 1997). As one might expect, the teachers also indicated that eliciting and noticing were easier to do in the classroom than taking action and responding! Any future teacher development would need to focus on the taking action and responding. And in doing so, would focus on the more risky and crucial aspects of the teacher's role and relationship with the students. It is the taking action and responding that determines whether the assessment is in fact formative or not. It is the taking of action and responding that gives the students feedback as to how to improve their learning.

The feedback obtained to-date suggests that other teachers and researchers are also interested in this clarification of the formative assessment process. The research findings lend themselves to the development of workshop materials for use in teacher education programmes to develop teachers' skills of formative assessment, both with respect to knowing about formative assessment and to being able to carry it out in the classroom.

In the next chapter, additional examples of formative assessment are given in the form of additional cameos to further illustrate the ten characteristics and model of formative assessment.

CHAPTER 6

CAMEOS OF FORMATIVE ASSESSMENT

In this chapter, a further six examples or cameos of formative assessment from the data are reported, having been selected to provide an illustration of the formative assessment observed within the classrooms in a holistic way, to highlight the complexity, and to illustrate the contextualised nature of the process. The cameos are considered to be episodes, where an episode is defined as all that happens from the time when the teacher started collecting the assessment information to when she or he had finished carrying out and evaluating her or his action. Each cameo starts with a brief description of the context of the episode. Then the researcher's representation of the teacher's story of the assessment episode, as obtained through the end-of-lesson interview, is presented. Next the researcher's field notes of the episode are presented to confirm and complement the interview data. In some cases, these add detail to the teacher's account. The researcher's analysis and interpretations of these episodes is presented last. Three cameos have already been given in chapter 3, the case study. The complete documentation of all thirteen cameos, as a part of the eight case studies, is given in the research report (Bell and Cowie, 1997, pp 48-245)

6.1   CAMEO: WASHING POWDERS

This episode is documented to illustrate 'noticing' as a part of interactive formative assessment. It occurred during a sorting exercise. Teacher 9 had provided the students with a set of statements about types of washing powder and washing conditions. They had to select those which would allow them to set up a controlled experiment to test the claim: 'Sprite washes cleanest in cold water and better than any powder cleaners if used in hot water'. The teacher had assumed incorrectly that the students understood the notion of 'control'. Teacher 9 said her awareness of the students not understanding the idea of a 'control' arose during a whole class discussion when she couldn't make sense of the students' comments. She described deliberately checking on individual students' understanding by going around and talking to them:

> I was going to check though really quickly that they had done the card exercise and
> .... knew what control was, but I found that they had interpreted the question very
> differently. Naturally the next thing to do is to see how many other people have got
> that same (idea) ...

> (Did it become more conscious?) Oh, yes. From that group onwards I was looking for
> where they were going. Like J, with the soluble thing. . ... (T9/D4/96)

[A full explanation of the data codes is given in the appendix].

The researcher observed this pattern on another occasions and discussed it with the teacher. Teacher 9 indicated this was a pattern which she often followed:

> (To) go around and see what everyone else had done. That is probably a pattern with me. It is matching. There's formative assessment. It is matching in my head where I think they're going and where I think, they think, they're going, too. I don't know that it's just what I think. It is also where they have given me the impression ... they are going. When I started talking about the control stuff it sounded like it was boring, they knew it. (T9/D 5/96)

The informal process of interactive formative assessment described here required teacher 9 to attend to student comments and questions. It required her to be sensitive to and to notice the implications of what the students were saying, in terms of the sense they were making relative to the science understandings she was specifically hoping to promote.

Teacher 9's descriptions of the lessons suggested she was very interested in and noticed students thinking, and was particularly sensitive to students' alternative conceptions. For example, she described the diversity in students' ideas (T9/D2, D3/96), how they transferred and linked ideas from one context to another (T9/D3, D11/96) and she referred to the scientifically unacceptable conceptions they held (T9/D4/96). Her noticing of students' thinking was consistent with her expressed interest in their thinking and desire to foster it within her classroom (T9/FN1/96; T9/D 1/96).

In addition, teacher 9 noticed the students' body language, the tone of their discussions and the comments they made about their own understanding (T9/EOY/96). For example, after one lesson she described her impression that the students were making links and feeling more confident:

> They were beginning to see the patterns and they were not as stressed. On Friday they were quite stressed about 'There is no pattern to this', 'I can't see where this is going'. When I went around (today) and looked at what they were doing, they were ok. ... (T9/D 10/96)

6.2   CAMEO: THE COLOURS IN BLACK INK

This cameo is presented to illustrate the process of interactive formative assessment. The teacher set the scene for this episode during the second lesson of the year. The purpose of the second and third lessons was to revise scientific ways of investigating and the teacher posed two questions for the students to solve: 'Which is the most common letter on a page of newspaper?' and 'Which black pen in this class contains the most colours?' Students worked on these questions in self-selected groups while they formulated and tested their hypotheses for each question. The students worked on the first question during the second lesson and completed their analysis for homework. They discussed their results and methods at the beginning of third lesson. This episode started when the teacher asked the students if they remembered the second question: 'Which black pen in this class contains the most colours in its ink?'

The researcher field noted this episode as:

The teacher told the students that she was using this activity were to revise the scientific ways of investigating and to provide an opportunity to assess the students' practical skills. She wrote the question on the board: 'Which black pen in this class contains the most colours?' She explained that to get the answer the question they would have to carry out an experiment. She told the students she could provide them with boiling tubes, filter paper and water to help them do this. She said the students' task was to design, carry out and evaluate an experiment to answer the question. One student asked if they should use more than one black pen. The teacher replied they should and asked for a show of hands from those students with black pens. A large number of students had black pens. The students moved into groups.

The teacher moved around the class talking to different groups. During this time the researcher observed one group of three students. These students and those around her discussed doing a similar experiment in the third form. They recalled they had used meths and a petri dish. The students in the group the researcher was observing, pooled their black pens and then two students discussed the method they should use while one student wrote in her book. This student and one of the others discussed their hypothesis. The three students recorded the same hypothesis and method in their books.

The teacher approached the observed group. One of the students read her hypothesis to the teacher. The teacher stated this was an acceptable scientific hypothesis. The teacher left.

One student collected a piece of filter paper and a petri dish. She and one of the other students folded the filter paper in half and placed dots of the different inks 2 cm up from the folded edge. All three students watched when the water was added. They decided that only the NY (a tradename) pen ink had separated and discussed the need for meths.

Two of the students moved to another group. One student repeated the experiment. The teacher approached her and she told the teacher that only the NY ink had more than one colour. The student looked more closely and then said one other ink showed some movement. The teacher asked the student if she thought the time she had waited or the solubility of the inks were factors which might have influenced the experiment.

The group sat together while they wrote up their conclusion which was that the NY ink contained the most colours.

Near the end of the lesson the teacher called the class back together. She held up three different experimental set-ups. She emphasised that all three used water, ink and paper and that these were essential. She concluded this episode by stating: 'I guess the answer to the question is that the NY ink has the most soluble colours'. (T9/FN 3/96)

The researcher discussed this episode with the teacher immediately after the lesson. Teacher 9 described her formative assessment as:

The formative thing was ... I didn't know they had already done chromatography because my third form hadn't done it. That brought out the extension conversation down the front here. ... I found it really interesting that when I gave them that gear they immediately came down to the front and said 'Where's the meths?' and 'Where's the petri dishes?'. ... the newspaper experiment was fresh ... the second (black pen) experiment was very coloured by their previous experience. ... That was unexpected. (T9/D 3/96)

Later in the discussion, the researcher asked teacher 9 about the group she had worked with:

R   Thinking about that group I ended up in, did you notice anything about them within this lesson?

T	That they had got themselves a range of things to work with. (The teacher asked the researcher if she had anything to do with this and the researcher stated that she hadn't) ... it turned out that their method was the most interesting compromise between what they knew they had to do and being innovative ... actually doing it another way. I found it fascinating that they borrowed the petri dish and used the boiling tube method.
R	Was there any formative assessment action for you, with them?
T	For them or for me?
R	For you with them as a team.
T	I didn't observe them as a team. I observed J doing most of the work. ... Most of my assessment today came from these front benches rather than down the back. ... I went around and saw what everyone was doing ... I saw that they (the innovative group) were transferring from here to their everyday life. ...
R	Was there any 'What shall I do?' or doing things today?'
T	Oh yes. There was choosing what the focus would be today ... I hadn't done that before for either of those experiments. ... as we went through, that the focus would be different methods rather than the best method. I felt the same with the chromatography .... If I was doing chromatography .. I would going into that solvent only works with these ones ...... so it appears that only one of these has any other colours in. Is that true or is there another story? ... if it was to do with pigments and colours ... I would ... (T9/D 3/96)

The researcher asked teacher 9 about a specific events:

R	I would like to ask you ... when you were talking to J, she said something like 'Only the NY has worked' and then she noticed the little bit of yellow coming off the top of another one. You talked to her and said 'that's if we're talking about time' and you mentioned the ink being soluble in water. I noticed that when you wrapped up you mentioned ... that perhaps the answer to the question was that we've got very few soluble colours.
T	Yes. This class, now I know that they've done that before and should be comfortable with the word soluble ... that should have created a picture in their minds that it's not what's there, it's whether or not it's dissolving. ... My question was 'Which had the most colours and we haven't really answered that question. They've found out we can answer things in different ways but that's a very false question ... I had not given them sufficient gear to answer the question ...
R	... when I was listening I interpreted that perhaps you assessed that J thought that only the NY moved and you wondered and fed back to her 'Was it to do with the water?'
T	It wasn't based just on J. It was based on the whole class. That's the first time that thought came up but as I went around I kept meeting it, I kept thinking... I do not want to give them the wrong message here. I do not want them to go away thinking that only the NY contains other colours ... which is what I was setting them up to do if I didn't say anything more. (T9/D 3/96)

This episode was presented in detail as it is interpreted as illustrating several features of the process of interactive formative assessment:

• teacher 9 gathered information from all the students as she had talked with all the students.

• teacher 9 considered she had detailed information on the thinking of only some of the students as she had spoken most to a group on the front bench.

• teacher 9 noticed those aspects the students' learning which she found unexpected. She had not expected the students to have done the black ink experiment before because she had not done the practical activity with her own third form in the previous year.

• teacher 9 noticed the students' thinking as it was represented by the diversity and ingenuity of their solutions to the problem. Her noticing of student thinking was consistent with her belief that thinking is important (T9/FN 1/96; T9/D 1/96).

• teacher 9 noticed and recognised a scientifically unacceptable idea (only NY black ink contained more than one colour). Enhancing the students' science knowledge was important to the teacher. Her interaction with one student, who held this idea, sensitised her to this view and so she was attuned to it in later interactions with the rest of the class.

• a scientifically unacceptable idea triggered her awareness of the process of interactive formative assessment. The teacher was concerned not to leave the students with scientifically unacceptable conceptions. It is possible that her interaction with a number of students, all with same idea, raised her awareness of this as an issue. Whichever was the case, teacher 9 was able to identify the student with whom she first became aware of this idea and to name the other groups who also held this idea. She described her awareness as culminating in an incident, in which a student concluded that black ink contained no colours.

• teacher 9 used information about what sense the students were making of the practical activity to refine her purpose for the activity. In this case, she told the students her purpose for the activity was to check they were able to generate an hypothesis, then design and carry out an experiment to test it (T9/FN 3/96). She said she considered the activity would provide her with an opportunity to assess their practical skills. After the lesson, she said that once she had seen the diversity of solutions the students had generated, she had decided 'the focus would be different methods rather than the best'. This focus enabled her to validate and encourage a diversity of thinking within the class, which was one of her stated long term goals. The interplay of her long term and short term goals was important on this occasion. Her formative assessment information enabled her to use the activity to promote one of her long term goals, within the framework of a short term goal. The flexibility of her purposes for the activity meant that the criteria she used to judge the students' thinking and actions were both pre-determined (could they use scientific ways of investigating - a short term goal) and emergent (being able to generate a number of solutions to a problem was valuable - a longer term goal associated with promoting student thinking and appreciating the limitations of a science activity)

• teacher 9 generated a number of possible actions, but she selected the one which she considered would most effectively meet her final goals for the lesson.

• She took generalised action with the class when she considered other students would benefit from feedback on a particular idea, and when she was not sure if she had provided every individual with the information during her interactions with them. Teacher 9 also acted with the class in an attempt to correct the students' perception that only one ink contained more than one colour.

6.3   CAMEO : MIXTURES

This cameo is included as it illustrates the process of interactive formative assessment. The lesson which is described was the first for the new term. Prior to the lesson, teacher 7 stated she intended to review the techniques used for separating mixtures, to get the students to complete a written task, and to evaporate a salt and water solution. She began by telling the class that separating mixtures was the last topic in the unit they had been working on the previous term and that once this was completed there would be a test. The researcher field noted the lesson:

> The teacher introduced the topic of separating mixtures and commented that 'some of these you have done before so we should be able to rattle through it'. She asked the students what they would do if she asked them to separate out the red smarties from a jar full of smarties. When the students stated they would pick them out on the basis of their colour she linked their comments to separating:

> 'the principle for separating components of mixtures .... is to find something different and use that property to separate them'.

> A textbook was distributed and the teacher read through and discussed the techniques of filtering, distillation, decanting, crystallising with the class.

> The first technique was filtering. The teacher commented to the class that they had already used this technique and asked if anyone could explain the principle of filtering. A student volunteered that 'when a filter had tiny holes only water and tiny objects can get through'. The teacher replied 'Exactly', restated this and linked filtering to sieving.

> The second technique was decanting. The teacher asked the class how many had decanters at home. One student asked what they were. The teacher described a wine decanter and how it worked. She stated decanting consisted of 'pouring liquid from the top of a solid'. She linked this with pouring the water from boiled potatoes.

> Crystallising, distilling and fractional distillation were covered in a similar manner. That is, the teacher described the technique, sought ideas from the students, answered their questions and made links to their everyday experiences.

> Next, the teacher introduced a 'thinking' task. She explained that she wanted the students to think about how they would separate out the two substances from each of the mixtures she was writing on the board.

> *How would you separate?*   *technique*   *property*

> *kidney beans from broad beans*

> *oil from water*

iron filings from sand

salt from sand

dirt from water

meths from water

gold specks from sand

She discussed what techniques could be used to separate kidney and broad beans.

The teacher moved around the class and spoke to a number of groups. She moved to the front of the class and said:

M has said 'What am I doing? I don't understand.' If she doesn't understand I am sure there will be lots of you who don't.

The teacher read out 'How do we separate kidney beans from broad beans?'. A student asked what a kidney bean was. The students and teacher discussed the shape of kidneys, and the shape and colour of both kidney beans and broad beans and linked these features to how the beans could be separated.

The teacher moved around the class talking to groups of students. One group checked that the two beans looked and tasted differently. A student asked if they had to use the techniques in the book. The teacher stopped the class and said: 'C. has made a good point,...'. She stated could use techniques other than those in the book.

A group asked the teacher if they could use filtering to separate oil and water. The teacher went to the prep room and returned with oil and filter paper. She poured oil on the filter paper and discussed whether filtering was appropriate. She moved to the front of the class and demonstrated the effect of oil on filter paper, saying 'If you are not sure of oil and filter paper ......' While she was doing this another group asked about sand and salt. She collected these from the prep room and invited the students to come and look at them so they could 'compare the size'.

She moved around the class talking to students. The teacher stopped that class and said:

A few things have become obvious ... you need to know the properties of the different things and you need to know what happens when you put them together. ... What about broad and kidney beans?

She discussed this example with the class and asked for suggestions about to how to separate oil and water. One student asked if she could decant the liquids. She explained that as oil floats on water, she would be able to pour the oil off first. A number of students queried whether the oil would be on top once the jug was tilted. The teacher drew two diagrams on the board.

One student suggested that the water would also come out. She explained her reasoning and at the teacher's invitation she modified the teacher's diagram. There was a general discussion of the shape and position of the boundary between oil and water when the jug was tilted. The teacher noted that the lesson was nearly at an end and concluded the discussion by saying: 'Maybe we need to try this ... tomorrow'.

She asked the students to complete the task for homework.

> As they left the laboratory the group beside the researcher asked the teacher what a technique was, whether they could use techniques not in the book and if they could separate gold and sand using a magnet. (T7/FN1/96)

Teacher 7 and the researcher discussed this lesson. Teacher 7 stated she had expected the students to find the principle of separating substances a simple topic:

> I expected that they would have that sort of down. I suppose I expected that this is a really simple topic that everyone would ... get just like that. (T7/D1/96)

Teacher 7 highlighted the fact that her expectations were not realised. She had expected the students to be familiar with the techniques for separating substances, the properties of the substances she had selected for them to separate, and the way these substances reacted when they were mixed. For example, she spoke of expecting the student to have had experience with filtering:

> I expected them to have done something like that before. They've said that. When ... we did filtering, some of them said, 'Oh we've done this'. ... some of them have done various topics. They've done water, they've done this, they've done solids, liquids and gases, according to them. ... they seem to touch ... on the ideas, but they obviously haven't gotten the ideas that I'm hoping to bring out of them. But they think they've done it because they've covered the topic before or they've mentioned these things before. (T7/D1/96)

The teacher spoke of being surprised that the students did not have a 'general knowledge' and experience of the substances she was asking them to separate:

> ... they didn't know what kidney beans and broad beans were. ... I thought they were going to be really obvious things to use. But they didn't actually know what they were. ... it became obvious that they didn't actually know enough about these things to be able to separate them. ... there were quite a lot (of students) who didn't know what oil did. To me it was obvious it was going to float but it wasn't to them because they didn't have the experience of that. (T7/D1/96)

The students' comments during her introduction of the principle of separation and the possible techniques for separating mixtures, had led the teacher to conclude that the students understood both the principle and its application within the different techniques. So their response to the activities was unexpected.

When asked what formative assessment she considered she had undertaken during the lesson, teacher 7 spoke of becoming aware of her use of formative assessment when she was introducing the task. Her awareness was triggered when she, unexpectedly, needed to ask a number of direct questions in order to obtain the information she needed:

> When I was trying to explain the idea ... why you can separate things using the different properties ... I was just aware of asking a lot questions. ... I was hoping more was going to come spontaneously. ...They were going to think, 'Done this. Know it' and it was going to come. (T7/D1/96)

Teacher 7 said how she had become aware that the students had a limited knowledge of the properties of the substances she had included in the mixtures, through their questions and comments:

> T   G's bald 'what on earth are kidney beans like?'. She had no idea. ...
> They were asking me, 'What's this like? What happens if this? ... they were asking me questions.

R   So it was lots of questions?

T   Yeah, ... just hearing the things that they were saying which were obvious nonsense. Like 'We could decant them, kidney beans and broad beans', ... things like that which meant they didn't understand the term. Or they hadn't looked. ... And they needed to come to grips with what the properties were. ... they were asking questions like 'What does oil do? 'What will oil do if you filter it?' (T7/D1/96)

The teacher identified the students' lack of prior knowledge as a misconception she had held. She said she had expected that 'they would know what salt and sand are like and be able to apply that theory'. As a consequence of what she found out, she said '... we needed to do something about the properties of salt and the properties of sand and then mix them and see' (T7/D1/96).

The actions the teacher took were to revisit the principles of separation, focusing on the use of a difference in the properties of the substances:

It became obvious that they didn't know. I had to show them what oil did when you put it on paper. Because they didn't realise it would soak in. ... they wanted to know if it would go through or if it would sit on the top. And I said you know what happens if you put oil into water. No. (T7/D1/96)

Teacher 7's actions are interpreted as reflecting a review of the time frame for her goals. She stated her initial intention was to move quickly through the task and focus on separating salt and water through evaporation. On finding out the students' level of prior knowledge, her goal became one of increasing the students' knowledge of the individual properties of the substances she wanted them to separate and the relevance of these to the mixture. She began the next lesson by demonstrating the separation of oil and water, the students then separated meths and water, and sand and iron filings. She also demonstrated the non-magnetic nature of gold.

During this lesson, the teacher's interaction with the students also 'required' her to generate additional goals. For example, the discussion on what happens when a mixture of oil and water are tilted, generated a high level of interest and diverse student opinion. Teacher 7 said 'I've got to show her ...' when she spoke about one student's view of what would happen when a mixture of oil and water is tilted. Clarifying this student's ideas became an additional goal for her. She addressed this questions with the whole class at the beginning of the next lesson. It is unknown if she would have acted with the whole class if there had not been general interest and discussion.

The main points illustrated by this cameo are:

• teacher 7 used interactive formative assessment (noticing, recognising and responding).

• teacher 7 became aware of undertaking interactive formative assessment when the unexpected nature of the students' responses to the task necessitated her asking a number of questions. Questions and suggestions from the students while she was moving around the groups alerted her to the nature of the students' scientifically unacceptable ideas.

• the identity of the students who asked questions was significant. Some of the students who asked questions were among those she considered 'thoughtful' and the most likely to understand (T7/D1/96).

• it was important that she was asked similar questions by more than one student, especially more than one thoughtful student. This helped focus her attention and raised her awareness of the problems. In this instance, the student question about kidney and broad beans would have been sufficient to alert her to the students' ideas.

• teacher 7 acted with both individuals and the class as a whole. For example, she showed a group the effect of oil on filter paper and then she demonstrated this to the whole class. This was a deliberate and considered action. She provided a number of reasons for this. She considered that the students who asked questions tended to be thoughtful, with a good understanding of ideas and said that if they were having problems, others would be too. She considered there were students who 'do not like to display their uncertainty to the teacher' (T7/D1/96). She said some students who knew they could not do the task, could not formulate a question to ask her in order to obtain help. She acted with the whole class to provide feedback to all these students.

• teacher 7 acted to address the students' scientifically unacceptable conceptions. She revisited the task requirements, the meaning of properties and techniques and she provided the materials for the students to look at. During the next lesson, she provided the students with the materials and they separated the mixtures themselves.

6.4   CAMEO: DENSITY AND THE TOWER

Within this cameo two episodes are compared and contrasted. The concept of density was the focus of both episodes. The first episode is described in the section called 'Density'. It illustrates the processes of planned and interactive formative assessment. The second episode is described in the section 'The tower' and is more summative in nature. The two episodes are compared and contrasted in the third section of this cameo.

Density

The episode described here took a whole lesson. The students had spent the previous two lessons doing practical work to explore the ideas of floating and sinking, density and the mass of air. They had spent time reviewing their conclusions to the practical activities. The teacher intended to discuss the students' ideas of density during the lesson. A brief summary of the researcher's field notes for the lesson is provided, followed by the teacher's and students' comments on the lesson.

> The teacher started the lesson by reminding the students they had started to talk about density during the previous lesson. She asked for someone to tell her what it was. A student said it was 'mass or volume'. The teacher rephrased this as: 'It is the mass of a <u>certain</u> volume'. She emphasised that the certain volume was important and recorded on the board:
>
> *density = mass of a certain volume*

She asked the students how they thought density, floating and sinking were linked. No one answered. She reminded them of the experiment in which they checked to see if cubes of different materials floated or sank. A student sitting beside her spoke to her. The teacher said:

T   Z has a thought. Z?

Z   All those which weighed less than water floated and all those which weighted more than water sank.

The teacher restated Z's idea and asked if someone could put it in a sentence using density. A student offered an answer:

S   Whether or not something floats depends on its density

The teacher asked if someone could provide another phrase which 'tells us a little bit more?' A student discussed lead floating and sinking. The teacher asked for a *general* statement. A student offered:

Things which are more dense than water sink and things that are less dense float.

The teacher asked the class:

Does that make sense to you all? If not put your hands up.

There were nods and 'Yeahs' around the room and she wrote on the board:

to compare the weight of materials we must use a fair test.

J gave the formula $p = m/V$.

She continued writing on the board:

The rule of floating and sinking is

If it is denser than water it _____

If is less dense it _____

She moved around the class and then added to the board:

If it is the same density it will

Commenting that the last question opened up a whole new area of floating and sinking and drawing a diagram on the board showing substances floating at different levels. A student asked about the last questions and she discussed displacement.

She said:

The next question is .... Why do hot air balloons float?

She asked for someone to answer, suggested she would pick someone at random and named N. He replied 'the gases they use are lighter than air'. The teacher asked what gases were used. Students introduced helium and hydrogen. The implications of using these gases were discussed. The teacher concluded balloonists use hot air as they are able to control the ascent.

She then asked 'What is inside the balloon?'. The teacher talked with some students and then moved to the front bench and said:

Don't forget about the other stuff.

She indicated that the student needed to remember about the mass of the basket and the burners. She drew particles in the balloon.

A student asked:

S   Why doesn't the room fly when we turn on the heater?

T   OK, good question.

The students offered suggestions such as the room was too heavy and stuck to the ground.

She posed a series of 'questions for experts'

What happens to air as you go up?.

Up high will you need to fire the burner more of less?

What difference will hot and cold days make?

She concluded the discussion by saying:

There have been lots of ideas from different people ...

some of you have still not spoken ...

but I think that you've got it OK ...

(T7/FN8/96)

Teacher 7 initiated this episode with a question - a planned formative assessment. It was followed by a pattern of question from the teacher or a student, discussion and the teacher recording a statement on the board. During this lesson, teacher 7's focus was on the concept of density but her plan for achieving students' understanding was flexible. She said she had planned 'to ask a few questions' (T7/D8/96). Her flexible plan allowed her to respond to student comments by appropriating their ideas and building on them, weaving them towards the learning she wanted. Hence, she undertook interactive formative assessment. The diverse range of comments and her appropriation of some of them provided all students with a range of different acceptable explanations for density. The students said they found it helpful to hear a range of explanations (SG71/L8/96). They said that the teacher's explicit validation of some of explanations by recording them on the board, helped them to confirm that they did understand (SG71/L10/96). This episode is viewed as an illustration of formative assessment in that the teacher provided the students with feedback on the appropriateness of their explanations. She also acted to provide opportunities for the students to test out their ideas. Some students took advantage of these and received individual feedback on their thinking in the whole class situation. By recording some ideas on the board, she provided explicit feedback to all the students on what counted as an acceptable scientific explanation of the concept of density.

In the next section, another episode in which the whole class focused on the concept of density is described.

The Tower

The episode described here took place two lessons after the episode discussed above as 'Density'. Teacher 7 had set up a measuring cylinder containing six liquids with a solid floating at each interface. At the end of the previous lesson, she had asked the students to explain 'What is happening here and why?' for homework. This episode started when she asked for students' answers to the question at the beginning of the next lesson. The researcher field-noted this episode (the students who were involved are coded S1, S2, S3 and S4):

> The teacher told the students she wanted to hear 'the views of those students who might not usually answer' and that she would randomly select some students. She did this by pointing at the roll and then naming a student. She asked four students for their answers. The first said:

S1   The top is less dense than the bottom ... in the middle it's the same density.

T   ... Can you explain the middle with the grape and the sea and usual water?

S1   Is the water different?... the grape is the same density as water and salt water. ... Are you trying to confuse me?

T   I am trying to understand how you think. If I ask a question ... I am trying to understand how you think ... to clarify the question ... I expected you to talk about the materials in there ... I am not sure you understand.

S1   The grape is less dense than sea water ... but not as dense as water is. Is that what you wanted?

T   That makes sense ... now I think you understand what I wanted.

> S1 was awkward and defensive during this sequence.

> The second student said:

S2   ... things denser than water sank ... things in the middle floated

T   ... Do you think the bung is sunk?

> S2 spoke very quietly and other comments were not heard. The third student said:

S3   I am not sure ... I guessed

T   That's OK

S3   I really have no idea. It might be because ... (she described the layers of
           liquid, then the materials using the words lighter)

T   Sounds to me you have got it really straight. Can you say that about
           aluminium again?

S3   Aluminium is heavier than ... but lighter than mercury so it stays above it.

T   Instead of lighter we can say denser.

The fourth student said he had not done his homework but he said:

S4   ... is heavier than water ... then there are grades of heavy ... salt water is
           not as dense as glycerine ....

T   OK ... good. What about the things?

S4   The material less dense than the substances are floating and those denser
           than the substances are on the top.

The teacher asked the students to write down the material in order of density. She
moved around and looked at their books and talked with some students. Some students
discussed what glycerine was. At this time C asked 'How dense are we?' so the class
heard. The teacher asked the class 'What makes people float and sink?'. The
students talked about breathing in and out while floating. The teacher said that 'we
are about the same density as water' and then asked:

T   Could we float on mercury? ... Hands up. Make a decision. You can't
           sit on the fence. ... Who thinks we would float? (most students put their
           hands up). Who thinks we would sink?

Only two students indicated they thought they would sink. She asked each student
'Why?'. They both responded quietly and their replies were not recorded. The
teacher did not press for further explanation. The teacher then asked a student to read
out what he had written down as the order. He read these out from heavy to light. The
teacher asked who had a different answer but no one responded. The teacher said the
substances could also be listed in the other order. A student suggested that the lightest
substance was air. The teacher replied 'Good one ....'.

The teacher stated she considered the students understood the idea of density and
introduced the next activity. (T7/FN 10/96)

The researcher considered the students were tense during this lesson, especially
while the teacher was selecting a student to answer. They appeared attentive while
other students were replying.   When they were writing down the order of the
substances, they talked quietly together. The teacher considered most students had
been able to order the substances (T7/D10/96).

The researcher interviewed S1 and S3 at the end of the lesson as they were
members of her interview groups.

After this lesson, teacher 7 described the episode as one of formative assessment.
She said she had focused on 'a smaller bit of knowledge'; she was:

'seeking confirmation the students understood what had gone before ... checking up
that they could transfer the ideas ... I expected them to have got it ... I hoped they

would use the right *language* to show they understood about different densities. ..I said 'Here's a situation, explain it' (T7/D10/96).

One student group indicated they were aware that the tower question had links with density but this was a new situation. They considered the teacher was interested in whether they could transfer ideas, saying they thought she wanted them to 'figure out' another situation:

S1   She hasn't really taught us. She has but ..

S6   She has but we still didn't. I suppose she was trying to see what knowledge we knew ourselves.

S3   What we knew.

S1   Without her teaching us.

...

S1   To see if we could do it ourselves.

S4   You said she has but. Did you mean she hasn't taught you that tower thing?

S1   She only..

S3   Explained it briefly.

S6   She just told us what's in it.

S5   But she wanted us to figure out. (SG71/L10/96)

Teacher 7 said she was surprised at the vagueness of the first student's response. She said :

... I expected them to be spot on, straight away (T7/D10/96)

In this case, teacher 7's expectations formed her pre-determined criteria for what counted as an acceptable response to her question. These criteria were implicit for she did not disclose them to the students until after the first student had answered her question. The first student had to predict what the teacher's criteria for a correct answer might be when formulating his answer. In her response to his answer, teacher 7 made her criteria for a successful or relevant answer more explicit. This effectively 'funnelled' (Bauersfeld, 1988) the student's subsequent answers so he was able to display the knowledge she was seeking. This procedure might be viewed as providing the student with the opportunity to display the knowledge the teacher was interested in. It might also be viewed as limiting the student's opportunity to display what he knew. Teacher 7's funnelling of the student's answers was consistent with her 'confirming' or 'checking' purpose for this activity and was therefore a convergent formative assessment. However, it was also effectively a summative assessment activity as it replaced the test at the end of the unit. Teacher 7 wanted to check the student had the scientific understanding she was interested in, not what he understood, which had been her focus in 'density' lesson. Towards the end of the discussion of the tower lesson, she described it as a 'summative activity which included formative assessment' (T7/D10/96). She explained that she was using the activity to check that the students understood the idea of density, that is, she was using it to sum up the students' learning. She said she was also responding to and trying to build on the answers she received. (T7/D10/96).

Comparing the density lesson and the tower lesson.

It is significant that the student's responded very differently to these two lessons during class time and during the interview, depending on whether the students felt the assessment was to be for formative or summative purposes. When teacher 7 compared the two lessons, she said that her intention in the first one had been to pull together three or four ideas. She indicated she considered all but two students did this. Her intention for the second lesson, was to 'check out' and 'confirm' the students understood density.

The students described the two lessons differently during the end-of-lesson interviews. During the interview after the tower episode, the students focused on being questioned, rather than on learning. They considered the teacher was expecting a specific response and viewed this episode as 'like the end' and 'this is the conclusion, sort of thing ... a summary'. They described the situation as 'harder'. They perceived the teacher's purpose had changed, for example, one student said:

> ... but when it is specific, then it is harder ... like that tower thing. We had to write down something. (SG71/L10/96)

The students viewed the tower episode as a summative assessment activity even though the teacher did not tell them it was. She only recognised the extent to which her purpose was summative towards the end of the discussion of the lesson. The targeting of students to answer a closed question, her indication that she was seeking a particular form of answer and her probing for this indicated to the students that this was a summative activity, that is, her intention was to sum up their learning. This was evident in the contrast between their views of the tower and the density lessons. They described the discussion on density as a time in which they were negotiating, with the teacher's mediation, an acceptable definition of density:

S6   Everyone in putting in their ideas. A discussion is different.

S3   Because everyone had different ideas....

S6   In a class discussion everyone puts forward their ideas. ... I do in class discussions but not in (SG71/L10/96)

During a report back interview two months later, the researcher described these two lessons to the students and asked them how they decided whether a teacher was checking on their ideas or discussing them with them. One of the groups differentiated between these two activities in terms of how the teacher asked questions. Of discussion they said:

S4   How do you work out the difference between the discussion time and the checking time?   What tells you?

S7   In discussion time it is not as much, she is asking you questions. You are giving more answers and she is not asking as many questions. You are giving her thoughts and stuff.....

S4   I have sort of got this but I still don't understand, I'm sorry. In discussion time it is different, she is asking fewer questions?

Ss   Yeah

S1   But not of, you have got to use your head

S7   She is giving more ideas ...

S4   She is asking fewer questions and they are, are they sort of different?

S8   Longer.

S7   Some are a bit longer than

S1   Quick questions

S7   If she says the question then someone ... will answer it. Then some else will give their view and someone else. (SG72/MC/96)

Of checking they said:

S4   In checking?

...

S1   Giving us a whole lot of original questions.

S8   ... she will ask one particular person.

S7   And she just goes (on) until someone gets it right, really.

S1   She just goes 'You', 'You', 'Yes'.

S7   And if it is right, she just goes onto the next one (question). (SG72/MC/96)

The students said the teacher asked longer questions which required them to think during discussions. They said they volunteered ideas and the teacher contributed more ideas during discussions. When talking about the teacher's actions when she was 'checking' on their understanding they described an 'initiate, respond, evaluate' sequence. They said she asked more and 'quicker' questions, naming students to respond until she got the 'right' answer. They indicated that they considered they were supposed know to the answers to these questions so they did not usually learn anything. One student said if he was unable to answer he listened to others' replies. One student commented that the teacher was also checking on their ideas during the discussion:

> She does both. ... I think the checking time cuts in with the discussion. (SG72/MC/96)

The comments made by these students suggested they were able to determine when a teacher's intention was to sum up their learning, that is, to find out if they understand, and when it is to inform their learning, that is to find out what and how they understand. Their comments suggested they were able to differentiate between formative and summative assessment occasions even if the teacher does not make this explicit. In this case, the students differentiated between the two on the basis of the teacher's questions and responses. What is critical in this distinction from the teacher's perspective, is that the students indicated they were reluctant to volunteer and reveal their thinking when they considered the teacher was checking on their understanding. The risks were too high.

In summary, these cameos further illustrate what formative assessment involves for both the teacher and students. It is a complex, highly skilled action that has both individual and social aspects. These cameos illustrate that formative assessment is a highly contextualised, purposeful, intentional, responsive, linguistic action by teachers and students and that it is integrated with teaching and learning.

CHAPTER 7

LEARNING AND FORMATIVE ASSESSMENT

In this chapter, we wish to build on the summaries and overviews given in chapters 4 and 5. We wish to theorise on our research findings, which suggest that formative assessment can be viewed as a purposeful, intentional, responsive activity involving meaning making; an integral part of teaching and learning; a situated and contextualised activity; a partnership between teacher and students; and involving the use of language to communicate meaning.

Our purpose for theorising about formative assessment, is to account for formative assessment within the current debates on cognition, learning and language in the classroom, rather than within the current debates on accountability in education. If formative assessment is considered to be a part of learning, then theorising about learning should be useful in theorising about formative assessment. Nuthall (1997), in his review of recent studies of student thinking in the classroom, identified three broad categories. Firstly, there are those studies (the cognitive constructivist perspective) that incorporate learning and thinking into a broad conception of cognition and students are seen as creating or constructing their own knowledge and skills. A second category (the sociocultural and community focused perspective) contains those studies that are primarily sociocultural in their orientation. Learning and thinking are seen as social processes or social practices, that is, practices occuring in social contexts, - between, rather than within, individuals. The third category (the language focused perspective) contains studies that have a primarily language or sociolinguistic orientation. 'Here, the language of the classroom is both the content and the medium of learning and thinking. What students acquire are the lingusitic "genres" of the disciplines' (Nuthall, 1997, p. 1). These three categories are evident in the following discussions of learning and formative assessment.

As discussed in chapter 2, previous debates on learning and science education, have involved discussions on behaviourist and constructivist views of learning. More recently, sociocultural views are being used to theorise learning in science. We argue here that sociocultural views of cognition and learning can be used to theorise about formative assessment and in this chapter we attempt to theorise the findings of the research project, using sociocultural views.

Within the science education, education and other literature, current debates on learning cluster around sociocultural views of cognition and learning which are variously described as social cognition (Resnick, 1991; Augoustinos and Walker, 1995; Salomon and Perkins, 1998); social constructivist view of learning (Bell and Gilbert, 1996; Driver, Asoko, Leach, Mortimer and Scott, 1994); situated learning

(Lave and Wenger, 1991; Hennessy, 1993); apprenticeship, guided participation, participatory appropriation (Rogoff, 1995); distributed cognition (Salomon, 1993b; Carr, 1998); mediated action (Vygotsky, 1978; Wertsch, 1991; Wertsch, del Río, Alvarez, 1995) and discursive activities (Harré and Gillett, 1994). These categories and associated descriptions are not mutually excluding, there is much overlap and lack of clarity. This is due to various categories having been developed from within different disciplines, for example, anthropology, sociology, psychology, and education. Different and similar words are used, different meanings are constructed and different emphases highlighted. In this chapter, we are not summarising these sociocultural views of learning per se, but rather we are theorising about formative assessment by drawing upon a number of sociocultural views of learning. In particular, we wish to account for the research findings and broadly theorise formative assessment as a purposeful, intentional activity involving meaning making; an integral part of teaching and learning; a situated and contextualised activity; a partnership between teacher and students; and involving the use of language to communicate meaning.

7.1   LEARNING AND FORMATIVE ASSESSMENT AS MEANING MAKING

For formative assessment to occur, students and teachers have to disclose to each other the meanings that they are making in the lesson, and negotiate a shared meaning. The feedback that the student receives about the 'gap' between her constructed meaning and the teacher's, will enable her to take action to bridge the 'gap'. The meanings constructed during formative assessment can be viewed as the mental representations of an object, event or idea, developed when an individual (such as a student or teacher) experiences and interacts with the environment. As a way of theorising about constructed meanings, constructivism has been a powerful and fruitful theoretical perspective in science education (Duit, 1994).

Taken in its most general form, constructivism asserts that all learning takes place when an individual constructs a mental representation of an object, event or idea. Mental representations are used as a basis for mental and physical action, and both enable and constrain an individual's process of meaning making (Resnick, 1991). A personal constructivist view of learning in science was developed in the 1980's (for example, Osborne and Wittrock, 1985; Osborne and Freyberg, 1985; Driver and Bell, 1986; Driver, 1989). The findings of this research suggest that students and teachers had to co-construct a shared understanding during the formative assessment process.

One of the main criticisms of personal constructivism is that this view of construction ignores the socially and historically situated nature of knowing. It gives 'primacy to abstract mental structures and rational thought processes at the expense of the historically and socially constituted subjectivity that learners bring to the reasoning process' (O'Loughlin, 1992, p. 800). In response, there has been a growing recognition of the role of the social and cultural aspects in learning in science as well as the personal, constructivist aspects, and science educators have sought to develop a social constructivist view of learning (Driver, Asoko, Leach, Mortimer and Scott, 1994; Bell and Gilbert, 1996). A social constructivist view of learning was proposed by Bell and Gilbert (1996, p. 50), recognising these components:

- Knowledge is constructed by people.
- The construction and reconstruction of knowledge is both personal and social.
- Personal construction of knowledge is socially mediated. Social construction of knowledge is personally mediated.
- Socially constructed knowledge is both the context for and the outcome of human social interaction. The social context is an integral part of the learning activity.
- Social interaction with others is a part of personal and social construction and reconstruction of knowledge.

Bell and Gilbert (1996) supported a view of learning (with respect to teacher development) which considered both the development of the individual's construction of meaning towards the socially agreed to knowledge and the reconstruction and transformation of the culture and social knowledge itself. In other words, such a view of learning would acknowledge the partially determining and partially determined characteristic of human agency - the interaction of the individual with the social can change both. The personal construction of knowledge was seen as mediated by socially constructed knowledge and the social construction of knowledge was seen as mediated by personally constructed knowledge. As indicated in chapters 3 and 4, formative assessment, as researched in this project, is a highly contextualised activity. The social context mediated the process of formative assessment.

In addition, in focusing on how students individually constructed understanding from experiences and interactions with the physical environment, personal constructivism was also criticised in that it did not address the affective and intentional aspects of thinking and learning (Pintrich, Marx, and Boyle, 1993; Gilbert, 1997, p.228) and that its epistemological basis was flawed (Osborne, 1996).

7.2   LEARNING AND FORMATIVE ASSSESSMENT AS SOCIOCULTURAL ACTIVITIES

Since the mid-1990s, there has been considerable theorising on learning with respect to sociocultural views of learning in education and in science education, with the goal of considering both individual and social aspects in the process of meaning making. In other words, the mental activities of individuals and an individual's meaning making are considered along with their socially and historically situatedness.

Salomon and Perkins (1998) in arguing the case for something called 'social learning', distinguished six meanings of social learning for the sake of conceptual clarity. The first three are of interest in the context of this chapter: i) socially mediated individual learning. For example, a teacher (the facilitating agent) teaches reading to a student (the individual learner); peer tutoring, collaborative, cooperative and reciprocal learning. This approach views the social system enhancing the individual's learning as an individual, striving to improve mastery of knowledge and skill.

ii) social mediation as participatory knowledge construction, that is, learning is the participation in a social process of knowledge construction. An example would be students participating in the construction and validation of scientific knowledge by the science community. In this view, it is impossible to examine mental processes independently of the sociocultural setting in which individuals and groups function (Wertsch, 1991). Hence, the study of learning must take into account the social contexts in which it occurs (Lave and Wenger, 1991 ; Resnick, 1991; Wertsch, del Río and Alvarez, 1995). It becomes unreasonable to separate cognition or motivation (or affect) from the socially mediating context or to separate individuals from their activities and the contexts in which they take place (Salomon and Perkins, 1998).

iii) social mediation by cultural scaffolding. That is, the learner is helped in some way by cultural artifacts, for example, tools such as computers, and sign systems such as speech genres.

These three meanings of 'social learning' underpin sociocultural views. In this chapter, sociocultural views of learning: situated learning, distributed cognition, and mediated action, will be used to theorise about formative assessment.

Learning and formative assessment as a situated activity

One sociocultural view of learning is that of learning as a situated activity. 'Situated activity' is a phrase used by Lave and Wenger (1991) to locate learning (and its formative assessment) in the processes of social interaction, not in the heads of individuals. In other words, learning (and formative assessment) is seen as a process that takes place in a co-participation framework, not in an individual mind. Lave and Wenger (1991) made a case for focussing on the relationship between learning and the social situations in which it occurs. Rather than defining learning in terms of acquisition or internalisation of structure, they viewed learning as the increased access of learners to participating roles in expert performances (Hanks, 1991). Learners are seen as 'participating in communities of practitioners and that mastery of knowledge and skill requires newcomers to move toward full participation in the sociocultural practices of a community' (Lave and Wenger, 1991, p. 29). Hence, learning is seen as a integral and inseparable aspect of social practice or, in other words, the process by which newcomers become part of a community of practice. Lave and Wenger (1991) used the term 'legitimate peripheral participation' to denote this learning through apprenticeship: 'legitimate peripheral participation is proposed as a descriptor of engagement in social practice that entails learning as an integral constituent' (p. 35). The novice comes to think and perceive as well as behave like the expert (Nuthall, 1997), in a process labeled 'appropriation' (Rogoff, 1993). Nuthall (1997) distinguished between the 'appropriation' of an expert's knowledge and skills from the concept of 'internalisation' that cognitivist theorists use to describe the acquisition of mental skills:

> Whereas internalisation refers to the incorporation of behaviour and knowledge into the cognitive processes of an individual mind, appropriation is the process by which two people come to understand each other and work effectively together. They each appropriate the product of their mutually evolving partnership in the activity. The process is inherently mutual, creative and situation specific (Rogoff, 1993) (Nuthall, 1997)

The term 'enculturation' is also used to describe the process:

> . . . learning is a process of enculturation or individual participation in socially organised practices, through which specialised local knowledge, rituals, practices, and vocabulary are developed. The foundation of actions in local interactions with the environment is ... the essential resource that makes knowledge possible and actions meaningful. (Hennessy, 1993, p. 2).

In other words, social processes can be seen as an integral part of cognition (Resnick, 1991), including formative assessment.

Learning and formative assessment as distributed cognition

Another sociocultural view of learning is that of learning as a distributed cognition and it is discussed here for its value in theorising about formative assessment. Situating cognition in social practice leads to a view of cognition as distributed across the context in which it occurs- hence the term 'distributed cognition'. When studied in real life situations (for example, in planning the family holiday), people appear to think in conjunction or partnership with others and with the help of cultural tools and artifacts (for example, language, maps and computers). A distributed view of cognition has been largely developed by those interested in the use of technology (for example, computers) in learning. Cultural tools and artifacts are used in cognition as a part of formative assessment, for example, language, computers, pens, paper.

A strong version of distributed cognition is that while 'individuals' cognition are not to be dismissed, cognition *in general* should be examined and conceived as principally distributed' (Salomon, 1993a, p. xv). The weaker version is that solo and distributed cognition are still able to be distinguished from each other and are taken to be in a dynamic relationship. It is this weaker version that is primarily discussed in this section, for Salomon (1993b, p. 113) reminds us that not '*all* cognitions, regardless of their inherent nature, are distributed *all the time*, by *all individuals* regardless of situation, purpose, proclivity or affordance'. Also, different writers in the field of distributed cognition, have differing views on the degree of distribution of cognition.

Hence, thinking can be considered to involve not just 'solo' cognitive activities but also distributed ones - distributed across other people and the sociocultural situation. Cognition (for example, learning and its formative assessment) is not seen as merely in-the-head activities, decontextualised tools and products of the mind (Salomon, 1993b). Nor is cognition seen as residing in the heads of individuals, with the social, cultural and technological factors relegated to the background. Distributed cognition can be summarised as referring to:

'1. the surround - the immediate physical and social resources outside the person - participates in cognition, not just as a source of input and a receiver of output, but as a vehicle of thought.

2. the residue left by thinking- what is learned- lingers not just in the mind of the learner, but in the arrangement of the surround as well . . .' (Perkins, 1993, p.90)

The social and artifactural surrounds, alleged to be 'outside' the individual's heads, are not only sources of stimulation and guidance but are actually 'vehicles of thought'. Distributed cognitions do not have a single locus 'inside' the individual. Rather they are said to be 'in between' and are jointly composed in a system that comprises an individual and peers, teachers or culturally provided tools. 'Distributed' is used in the sense of 'stretched over' (Salomon, 1993b) rather than just divided up. While not all cases of distributed cognition can be viewed as the same, they are seen as having one important quality: 'the product of the intellectual partnership that results from the distribution of cognitions across individuals or between individuals and cultural artifacts is a joint one: it cannot be attributed solely to one or another partner' (Salomon, 1993b, p. 112).

Our environment provides social, physical and artifactural support for cognition. Artifacts that help us think may be tools such as calculators, computers; symbolic representations such as language, mathematical symbols, graphs, diagrams; or the physical environment, such as work benches (Pea, 1993). Human cognition can be seen as distributed 'beyond the compass of the organism proper in several ways: by involving other persons, relying on symbolic media, and exploiting the environment and artifacts' (Perkins, 1993, p. 89). The social, artifactural and physical support in the surrounds can enable a person to deal with complex concepts that would be unmanageable for one person.

Cognition is also shaped by the situation with respect to affordances (Pea, 1993). Some technology affords greater opportunity for higher order kinds of thinking and learning (Perkins, 1993; Carr, 1998). 'Affordance' refers to 'the perceived and actual properties of a thing, primarily those functional properties that determine just how the thing could possibly be used. Less technically, a doorknob *is for* turning, a wagon handle *is for* pulling' (Pea, 1993, p. 51). In the educational setting, we hope that we can get a learner to attend to the relevant properties of the environment or object or text, such that the learner can join in. There will be variation in the ease with which a social, cultural, technological or environmental tool can be conveyed to and used by a learner in activities which contribute to distributed cognition.

In summary, distributed cognition is one way of viewing socially constructed learning and knowledge - knowledge that is 'socially constructed, through collaborative efforts toward shared objectives or by dialogues and challenges brought about by differences in persons' perspectives' (Pea, 1993, p. 48). Distributed cognition is also one way of viewing formative assesment. The definition of formative assessment given earlier in chapter 2, p. 11, can be viewed as being included in the above description of distributed cognition.

Learning and formative assessment as a mediated action

A third sociocultural view of learning and cognition (including formative assessment) is that they are mediated actions (Vygotsky,1978; Wertsch, 1991). A mediated action is a human action that employs mediational means, such as technical tools –for example, a computer, and psychological tools – for example, signs such as languages. As with situated and distributed views of cognition and learning (and its formative assessment), the focus is on human action in context. Again, the basic goal of this (and the other) sociocultural approaches to the mind is to create an account of human mental processes that recognise the essential relationships between these mental processes and their social, cultural and institutional settings (Wertsch, 1991).

A fundamental assumption of this sociocultural approach to the mind is that the unit of description and analysis is 'human action'. To understand mental functioning, one cannot begin with the environment (as in a behaviourist approach) or a human agent in isolation from the sociocultural settings (as in the cognitivist approach). Instead, a sociocultural view assumes that the notion of mental functioning is not limited to processes of the brain of the individual, and that it can be applied to social as well as individual forms of activity. In a sociocultural view of the mind, what is discussed and explained is human action and interaction. Wertsch (1991) asserts that in studying human action, one sees a close relationship between social communicative processes and individual psychological processes. Hence, to understand the individual, it is necessary to understand the social relations in which the individual exists.

'Mediated action' is a term used by Wertsch (1991) to emphasise that human action typically employs 'mediational means' such as tools and signs (including language). He gives support to Vygotsky's (1978) claim that the higher mental functioning and human action in general are mediated by tools (or technical tools) and signs (or psychological tools). For example, he cites the psychological tools of language, counting systems, mnemonic techniques, algebraic systems, writing, diagrams, maps and sees them as being part of the tool kit available to humans in the meaning-making process. A defining property of higher mental functioning is the fact that it is mediated by tools and by sign systems such as natural language.

The incorporation of the mediational means does not simply facilitate action that could have occurred without them (Wertsch, 1991); instead, as Vygotsky (1978) noted, by being included in the process of behaviour, the psychological tool alters the entire flow and structure of mental functions. Hence the agent and the means become inseparable. 'The action and the mediational means are mutually determining' (Wertsch, 1991, p. 119). For example, Wertsch (1991) gives the example of the blind person's stick. The stick is a particular shape and colour due to its use by a blind person. One cannot separate the stick and the blind person to make sense of it.

In summary, mediated action rests on assumptions about the close relationship between social communicative processes and individual psychological processes. The processes and structures of mediation provide a crucial link between historical, cultural and institutional contexts and mental functioning. It is the sociocultural situatedness of mediated action that provides this essential link between the cultural, historical and institutional setting on one hand and the mental functioning of the individual (for both learning and formative assessment) on the other.

Summary

The above sociocultural views of learning and thinking – situated, distributed and mediated action- all have in common, aspects that are useful in theorising about formative assessment. These are now summarised.

The main goal of a sociocultural view of learning, thinking and the mind is to create an account of human mental processes that recognise the essential relationships between mental processes and their social, cultural and institutional settings (Wertsch, 1991). In terms of the classroom, the goal is to account for the way social practices, including language, determine how and what children think and learn. Sociocultural views of learning (and its formative assessment) inform us that it is the whole of what goes on in classrooms that determines the learning, not just what is happening inside an individual's head. Overall the:

> 'sociocultural perspective was developed from a desire to see school learning within a larger cultural context. This led to a focus on the culturally embedded nature of the classroom processes and the central role that cultural norms and artefacts play in structuring the learning and the way we view learning' (Nuthall, 1997).

Likewise, formative assessment is a highly contextualised and situated activity. In understanding formative assessment, we need to consider not just the meaning making by an individual, but the context in which it is occurring.

Secondly, meaning is central to a sociocultural approach to mind. A sociocultural view emphasises the 'mind' rather than the 'brain'. If thinking is viewed as situated, distributed, or mediated action, then the mind is more than cognition or brain processing. It includes a wide range of psychological phenomena, such as, mental processes, self, emotions, intentions. Mind, in this view, goes beyond the skin and so we call it 'socially distributed' as mind and mediated action cannot be tied to an individual acting in vacuo. Mind, as it is used by Wertsch (1991), is defined in terms of its inherently social and mediational properties. When theorising about formative assessment, it is useful to consider the mind rather than just the workings of the brain. Emotions and intentions are as much a part of formative assessment as cognition to co-construct meanings.

Thirdly, sociocultural approaches consider both the individual and the social aspects of learning and thinking, given that the goal of a sociocultural approach to learning is to 'explicate the relationships between human mental functioning, on one hand, and cultural, institutional and historical situations in which this functioning occurs, on the other' (Wertsch, del Rio, Alvarez, 1995, p. 3). There is a need for such a sociocultural view as previous views of learning saw the learner as internalising knowledge, whether 'discovered', 'transmitted' or 'experienced in interaction' or 'constructed' (Lave and Wenger, 1991). These previous views established a dichotomy between inside and outside, and between the individual and the social, especially in individualistic, reductionism psychological debates. In contrast, sociocultural views focus on the 'mind' (rather than just the 'brain'); human action (rather than behaviour); and meaning making (rather than linguistic structure or mental/conceptual representation) (Bruner, 1990; Wertsch, 1991; Wertsch, del Río, Alvarez, 1995). Sociocultural views of learning then specifically address the issue of the distinction between the 'individual' and the 'social' in past psychological debates. Learning (and formative assessment) is seen as involving both individual and social aspects.

For example, Cobb (1994) asserted that ' mathematical learning should be viewed as both a process of active individual construction and a process of enculturation into the mathematical practices of wider society' (p. 13) - the description could also be applied to learning science. Cobb views the two perspectives - constructivism and the sociocultural as each telling half the story. Each perspective implies the other but foregrounds one aspect only. Salomon and Perkins (1998) also see the need to consider both, stating that one cannot be reduced to the other. Rogoff (1995) addressed the social and individual issue by proposing a sociocultural approach 'that involves observation of development in three planes of analysis corresponding to personal, interpersonal, and community processes' (p. 139). These are described as 'inseparable, mutually constituting planes comprising activities that can become the focus of analysis at different times but with the others necessarily remaining in the background of the analysis' (p. 139). One can become foregrounded, whilst the other two are not ignored, but backgrounded. She asserted that the development of children (for example) occurs through a process of participation and collaboration in social activities. These social activities can be in personal, interpersonal and community processes. The use of activities as the unit of analysis enables the social, the individual, and cultural environments to be described in relation to each other, for none is seen to exist separately (Rogoff, 1995). The activities which she focuses on in each plane are : apprenticeship in the community plane; guided participation in the interpersonal plane; and participatory appropriation in the individual plane. Hence, sociocultural perspectives on human functioning emphasise the *relationship* between mental processes and the sociocultural setting.

Salomon (1993b), in the debate on the relationship between and the relative roles of the individual and distributed cognitions, proposed a model for the interaction between individual and distributed cognitions. He described the components as interacting with each other in a 'spiral-like fashion, whereby individuals' inputs, through their collaborative activities, affect the nature of the joint, distributed system, which in turn affects their cognitions such that their subsequent participation is altered, resulting in altered joint performances and products' (Salomon, 1993b, p. 122). This spiral-like development allows for distributed cognitions and one's own 'solo' competencies to be reciprocally developed. Hence, the relationship will develop over time.

This position of the sociocultural views of learning, that are accepting of both social and individual learning and that differentiate between thinking and language, is appealing to theorising on formative assessment because the teachers and learners do attend to the social aspects of learning in the classroom, even though the education system as a whole (and in particular, assessment) focuses on the individual.

The fourth aspect in common between the three sociocultural views of learning is that of the methodological concern of the unit of analysis. One way to study both aspects (of individual and social aspects of learning and formative assessment), is to adopt the unit of analysis of human action (Wertsch, 1991), rather than focussing on the unit of analysis of concepts, linguistic and knowledge structures, attitudes, as often found in psychology, although they might be used in an analysis of human action. In his analysis, Wertsch (1991) sees mediated action as the irreducible unit of analysis and the person-acting-with-mediational-means as the irreducible agent involved. In a similar way, distributed cognition recognises that some activities are so highly contextualised, and dependant on the situation, that we cannot easily make the distinction between cognitive knowledge and skills, the context and the activity a person is engaged in. In effect, the unit of analysis for research and theorising on learning has changed from the individual alone to the individual plus those parts of the surround that may be supporting the cognition. Or as Perkins (1993) described it, the unit of analysis has changed from the 'person-solo' to the 'person-plus (the surroundings)'. Likewise, Pea (1993) takes the 'person-in-action' as the unit of analysis. That is, the unit of analysis is the person plus the 'resources that shape and enable activity are distributed in configuration across people, environments and situations' (Pea, 1993, p. 50). In other words, cognition (for learning and formative assessment) emerges or is accomplished, rather than being possessed. In using the 'person-in-action' as the unit of analysis, we need to consider the role of intent, desire and conation which shapes both their interpretation and use of resources for the activity. In this way, sociocultural views of learning address the integration of cognition, affect and conation, in a way that constructivist approaches do not.

7.3   LEARNING AND FORMATIVE ASSESSMENT AS A DISCURSIVE PRACTICE.

Another aspect of formative assessment is the central role played by language in meaning making; the partnership between the teacher and students, and communication. The role of language is theorised by a sociocultural view of learning and mind that considers learning as a discursive practice. Discursive views of learning, in contrast to the other three sociocultural views already discussed, theorise on only the social aspects of learning and give no functional value to a consideration of the individual aspects (Bell, in press). If we view learning (and formative assessment) as discursive activities, we are predominantly giving attention to language-in-use. In the classroom setting, we are giving focus to the ways language is used to promote thinking, learning and formative assessment. . The role of language (and other symbol use) is central to discursive psychology, in which we are examining human functioning in actual social and cultural settings. A discursive activity or practice is:

> ..the repeated and orderly use of some sign system, where these uses are intentional, that is, directed at or to something ...Discursive activities are always subject to standards of correctness and incorrectness. These standards can be expressed in terms of rules. Therefore, a discursive practice is the use of a sign system, for which there are norms of right and wrong use, and the signs concern or are directed at various things. (Harré and Gillett, 1994, p. 28 -29).

In short, a discursive activity is an intentional, normative action, using sign systems. The focus of discursive psychology is what talk and writing is being used to do, that is, what language is being used to achieve, rather than language been seen as an abstract tool to state or describe things. Language is seen as functional and used by people for example, to justify, explain, blame, excuse, persuade and present an argument. Hence, the notion of language-in-use relates to that of communication.

As with the other sociocultural views of learning, meaning is central when considering learning and formative assessment as discursive practices. Knowing what a situation means to a person, means we are able to understand what that person is doing (Harré and Gillett, 1994). We understand the behaviour of an individual when we grasp the meanings that are informing a person's activity:

> [Wittgenstein] came to realise that understanding and the phenomena of meaning or intentionality in general could only be approached by looking at what people actually do with word patterns and other word signs. He formulated the doctrine that meaning is the use to which we put our signs. He studied the use of language in 'language games', by which he meant complex activities involving both the use of language and the use of physical tools and actions, where they are ordinarily encountered... [he] came to see that mental activity is not essentially a Cartesian or inner set of processes but a range of moves or techniques defined against a background of human activity and governed by informal rules. These rules, unlike the rules-laws at work in supposed inner, cognitive processes, were the rules that people actually followed. They are most evident when we consider the correct and incorrect ways of using words....This understanding of human activity requires us to interpret the behaviour of another according to some appreciation of the self-positioning of the subject within the complex structure of rules and practices, within which that individual moves. (Harré and Gillett, 1994, p. 19 - 20)

Language and other semiotic (sign) systems play an important part in producing meaning, especially meaning as it shapes human action (Wertsch, 1991). Meaning here is viewed as being produced only in a social setting, and as a process, not a fixed entity inherent in a linguistic package:

> Wittgenstein emphasised the interactive and conventional nature of language. As a social practice, language has no fixed meaning outside the context in which it is used. Our perception of the world is shaped by the language we use to describe it: objects, activities and categories derive their epistemological status from the definitions we create for them. Within this view, thought and language are no longer separated. 'When we think in language, there are not "meanings" going through our mind in addition to verbal expressions. The language itself is the vehicle of thought (Augoustinos and Walker, 1995, p. 264).

It is usual to think of concepts as the basis of meaning, understanding and thinking. But concepts are expressed by words and words are located in languages. Thus, the discourses constructed jointly by persons and within sociocultural groups become an important part of the framework of interpretation and meaning. The communicability of thoughts is secured by the mutual intelligibility of a shared symbolic system, such as a common language (Harré and Gillett, 1994). The grasp of the use of a word/concept is an active discursive skill, rather than an inner cognitive skill, and learning is seen as the increasingly skilled use of social practices.

Therefore, the discursive view differs from the other sociocultural views in its non-mentalistic view of 'cogniton' and 'mind'. If priority is given to languages in defining what are psychological phenomena, then to present and understand cognition, it must be done in terms of the ordinary languages through which we think, rather than looking for abstract representations of them (Lemke, 1990; Harré and Gillett, 1994). Discursive psychology considers thinking, not as a mental activity, but as the activity of operating signs (for example, language). Hence, discursive approaches to thinking and learning differ from the sociocultural views in that they see no distinction between thoughts and language. Discursive approaches see problems in the assumptions that cognitive phenomena such as 'attitudes', 'emotions', 'categories', can be identified and located in an internal cognitive world – inside the head. Attention is given to the discourse itself and not the assumed underlying internal, static, mental states and processes. Instead, discursive psychology is more interested in how people discursively constitute psychological phenomena to do certain things. Psychological phenomena are 'discursive actions' which are 'actively constructed in discourse, for rhetorical ends' (Wetherell and Potter, 1992, p. 77). The social processes are the cognitive processes. For example, categorization, as a psychological phenomena, is 'something we do, in talk', in order to accomplish social actions (persuasions, blamings, denial, refutations, accusations) (Edwards, 1991, p. 94). Because 'some constructions are so familiar, pervasive and common-sensical that they give an effect of realism or fact. People therefore come to regard some constructions not as versions of reality, but as direct representations of reality itself.' (Augoustinos and Walker, 1995, p. 269). Any internal cognitive realm is conceptualised as a form of situated practice. There is no notion of internal representation or model to assume cognitive mediation. (Augoustinos and Walker, 1995). Discursive (as post-structuralist) psychology is critical of social cognitive concepts such as representations, schemas, attitudes, categories which are hypothesised to be stable mental categories located within the mind. The position taken by Lemke (1990) and O'Loughlin (1992) is to deny the functional significance of individual mental processes (Nuthall, 1997). Their position is a relativist one, rather than realist, and as such may be unacceptable to many in the science education community. However, their denial is not a denial of the 'reality' of mind and cognition so much as a denial of the value of talk about mind and cognition (Nuthall, 1997).

Hence, a discursive approach questions the notion of a knowable reality by emphasising the socio-historical and political nature of all knowledge claims. Such a post-structuralist view of psychology stresses that words do not have independent, objective meaning outside the social and relational context in which they are used:

> Language is viewed as reflexive and contextual, constructing the very nature of the objects and events as they are talked about. This emphasises the constructive nature and role of language. ... As people are engaged in conversation with others, they construct and negotiate meanings, or the very 'reality' which they are talking about (Augoustinos and Walker, 1995, p. 266).

People live in two worlds: the physical world and the symbolic world. The physical or material world, is structured by causal processes. The symbolic world (the world of symbols) is organised by the norms and conventions of correct symbol use. It comes into being through intentional action. The relationship of a person to both these worlds can be understood through the idea of skilful action (Harré and Gillett, 1994), using complimentary manual and discursive skills. To operate in the physical world, we use manual skills. To operate in the world of symbols, we need to be adept at using discursive skills. As Harré and Gillett explain:

> There could not be a world of symbols unless there was a material world. But these two realms do not reduce to each other. We cannot explain the world of symbols and how it works by reference to physical processes... there could not be language and discursive processes unless there were brains buzzing with electrical and chemical processes and there were vibrations in the air and marks on paper. But those vibrations and those marks and buzzings do not constitute the mind. They cannot explain the intentional character of symbol use and the normative constraints under which symbols must be used. A buzzing in the brain cannot be correct or incorrect. It can only be. (Harré and Gillett, 1994, p.100)

A discursive approach to learning (and formative assessment) enables all three aspects of 'mind' (cognition, affect and conation) to be taken into account, rather than each being studied in isolation. A discursive view of mind asserts that to understand the mind is to study social interaction, not the biological brain operation of an individual. Harré and Gillett (1994) state that we need to move away from a focus on the individual as a rational subject and to look at a broader framework to understand meaning and rule-following.

To add emphasis to the notion that communication, mental processes, and conation are linked, Wertsch (1991) uses the notion of 'voice' (after Bakhtin, for example, 1986) meaning the speaking personality, the speaking consciousness. The notion of 'voice' is concerned with the wider issues of a speaking subject's perspective, conceptual horizons, intentions and world view. It always exists in a social milieu, that is, not in isolation from other voices. Voices produce utterances - a notion used by Bakhtin to focus on the situated action of language-in-use, rather than on objects that can be derived from linguistic analytic abstractions. Bakhtin's notion of utterance is linked with that of voice as an utterance can only be produced by a voice.

Considering how voices engage with one another is important to a discursive view of mind (Wertsch, 1991) for it is only when two or more voices come into contact (for example, when the voice of a listener responds to the voice of a speaker) that meaning comes into existence. And during formative assessment, the teacher and students share their meaning making and respond through their actions to improve learning. Taking into account both voices, reflects a concern for addressivity - the quality of turning to someone else. In the absence of addressivity, an utterance does not exist. Addressivity is not inherent in the unit of language (for example, word or sentence) but in the utterance. The notion of addressivity means that 'utterances are not indifferent to one another, and are not self-sufficient; they are aware of and mutually reflect one another' (Bakhtin, 1986, p.91 as quoted in Wertsch, 1991, p. 52)

Therefore, utterances involve both a concern with who is doing the speaking and a concern with who is being addressed. A teacher in giving feedback to a student about their learning, is concerned about speaking the voice of the scientist and how to phrase it for a learner of science. Utterances are inherently associated with at least two voices - the speaking voice may indicate an awareness of the addresse's voice. Bakhtin's concept of 'dialogicality', meaning more than one voice, is useful to Wertsch (1991). Human communicative and psychological processes are said to be characterised by a dialogicality of voices. That is, when a speaker produces an utterance, at least two voices are heard simultaneously. If human communication is characterised by a dialogicality of voices, then understanding is dialogic in nature. That is, to understand another's utterance is to orientate oneself with respect to it. There are different sorts of dialogues: face-to-face, inner dialogue, parody, and social languages within a single national language. Dialogicality is illustrated in the work of Scott (1997, 1998, 1999) who analysed classroom talk in terms of authoritative and dialogic nature of the discourse in the classroom. Authoritative functions of discourse are those that convey information and which emphasise the transmissive function of teacher talk. The dialogic function of teacher talk is that which the teacher encourages students to put forward their ideas, to explore and debate points of view. In a classroom, both functions of discourse are realised - the discourse has functional dualism. The situation is dynamic as the discourse shifts between authoritative and dialogic functions. Scott (1999) suggests 'that individual student learning in the classroom is enhanced through achieving some kind of balance between presenting information and allowing opportunities for exploration of ideas', that is, a balance between the authoritative and dialogic functions of the discourse.

Bakhtin (1986) also made a distinction between social languages (for example, teen speak) meaning discourses specific to a given social system at a given time, and national languages, for example, English, French. Another notion used by Bakhtin was the notion of 'speech genre', giving examples such as military commands, everyday greetings and farewells. He saw the 'speech genre not as a form of language but as a typical form [a type] of utterance' (Wertsch, 1991, p. 61). Wertsch distinguished between social languages and speech genres in that social languages relate to the different social groupings, whereas speech genres relate to 'typical situations of speech communication' (p. 61).

Social languages and speech genres are good candidates for the tools in the tool kit of mediational means for meaning making, for it is through these that utterances take on meaning (Werstch, 1991). Therefore, Wertsch would view social languages and especially speech genre, as a mediational means or tool for thinking and communication. Speech genres are seen to provide a crucial link between psychological processes as they currently exist and their cultural, historical and institutional settings. In classroom talking, the voices appropriated by the children can be fully interpreted only if one goes beyond the individual speakers involved. In order to interpret what it is that they have said and to identify 'who' it is that is doing the talking, one must look to the speech genre appropriated in speakers utterances.

Social languages and speech genres (as mediational means) appear to be hierarchically used. Wertsch (1991) uses the term of 'priviledging' to refer to 'the fact that one mediational means, such as a social language or genre can be viewed as being more appropriate or efficacious than others in a particular social setting'(p. 124). For example, in the science lesson, 'curriculum science' is priviledged over 'children's science'. Formative assessment plays a role in giving students feedback as to acceptable social languages and speech genre. And as Nuthall (1997) states, 'classrooms are language communities that develop their own forms of language'.

In summary, the mind, in a discursive approach, is seen as a social practice, rather than something to be sealed into its own individual and self-contained subjectivity. It is seen as a domain of skills and techniques that renders the world meaningful to the individual, rather than just the biological brain operation of an individual. The whole point of discursive psychology is to get away from 'mythical' mental activities (Harré and Gillett, 1994), the mind being considered as a non-mentalistic entity. This position is in contrast to the behaviourist tradition which views the mind as a private area, not available as a source of data. It is also in contrast to cognitive psychology which has a view of mental mechanisms and the existence of inner mental states and processes such as rule-following (for example, the scripts of Shank and the grammars of Chomsky). Cognitive science assumes that aspects of cognition (sensation, perception, imagery, retention, recall, problem solving, thinking) are mental entities, that is, have substance. Discursive psychology is not interested in mental representations but in meanings. Thoughts are not seen as objects in the mind but the activity and essence of mind. Thoughts reside in the uses we make of public and private systems of signs. To be able to think is to be a skilled user of these sign systems, that is, capable of using them correctly. Whilst, the usual meaning of 'cognition' is pertaining to thought, Harré and Gillet (1994) have found it useful to redefine cognition as pertaining to brain processes only. Hence, in their view, the study of mental processes (such as learning and formative assessment) can be seen as the study of discursive practices, rather than the study of internal brain functioning.

One of the main criticisms of a discursive approach is its inability to explain retention and memory, for it does not focus on mental activity (Augoustinos and Walker, 1995). In highlighting or foregrounding the social, the individual mental aspects are hidden or backgrounded. While teachers and students may attend to the social situatedness of formative assessment in the classroom, assessment regimes in most education systems focus on individual achievement and cognitive development. However, Harré and Gillet (1994) explicate a neural network model of mental activity, which they assert accounts for both the abstract representations of structures and functioning in the brain and nervous systems, and the metaphorical presentations of the 'grammatical' structure and relationships of intended, goal directed and norm-constrained human action (that is, discursive activities). In addition, Augoustinos and Walker (1995) argue it would be difficult to deny that cognition is taking place, for example, in reflection, learning, and deductive reasoning

A strength of a discursive view of learning and meaning is that it allows us to more readily, than the sociocultural views, give consideration to the notion of 'power'. If language is seen as a form of social practice and if meanings are seen as socially constructed, then what counts as coherent or meaningful depends very much on the power relationships, rather than on an absolute truth (Drewery and Winslade, 1997). Seen in this context, power is not the 'possession' of particular persons but is constituted in positions occupied by subjects in discourses. 'Discourse' is taken here to mean 'a set of ideas embodied as structuring statements that underlie and give meaning to social practices' (Monk, Winslade, Crocket and Epston, 1997, p. 302). This is important in assessment tasks where the power relations between teacher and student are influential on pedagogical and learning outcomes, such as in the issue of student disclosure, as discussed in section 4.3. Foucault's notion of surveillance and other techniques of power (Gore, 1998) might be a useful start from which to theorise about power in formative assessment as a subtle form of social control.

In conclusion, if formative assessment is theorised in terms of a sociocultural view of mind, the implications (Gipps, 1999) include that:

• formative assessment can only be fully understood if the social, cultural and political contexts in the classroom are taken into account

• the practices of formative assessment reflect the values, culture of the classroom, and in particular, those of the teacher

• formative assessment is a social practice, constructed within social and cultural norms of the classroom

• what is assessed is what is socially and culturally valued

• the cultural and social knowledge of the teacher and students will mediate their responses to assessment

• formative assessments are value-laden and socially constructed

• a distinction needs to be made between what a student can typically do (without mediational tools) and best performance (with the use of mediational tools).

• formative assessments need to give feedback to students on the assessment process itself to enable them to do self and peer formative assessment

• teachers and students negotiate the process of assessment to be used, the criteria for achievement, and what counts as acceptable knowledge.

The implications for viewing formative assessment as a discursive practice can be seen as:

• the relationship between the teacher and student is seen as important if the issue of power is to be acknowledged.

• the teacher's use of power be shared and negotiated in the assessment task so that she or he uses 'power with', not 'power over', the student (Gipps, 1999).

• the language of formative assessment, and the language of science are seen as determining of the learning outcomes in the science lesson.

• students understanding the use of language in the science classroom to make assertions, argue a case, or demonstrate supporting evidence (Driver and Newton, 1997).

Throughout this book, we have described, modelled and theorised formative assessment as a purposeful, intentional, responsive activity involving meaning making and giving feedback to students and teachers, to improve learning. We have viewed it as an integral part of teaching and learning; a situated and contextualised activity; a partnership between teacher and students; and involving the use of language to communicate meaning.

CHAPTER 8

DOING FORMATIVE ASSESSMENT

Black and Wiliam (1998) concluded in their review of the relevant literature, that the practice of formative assessment does improve learning. But what have the findings of the research reported in this book, got to say to teachers and teacher educators about improving the practice of formative assessment? In this chapter, we wish to consider the research findings in terms of what teachers (of science) can do differently in the classroom and what teacher educators can do to promote the professional development of teachers in the area of formative assessment.

8.1 CHANGING CLASSROOM ACTIVITIES, ACTIONS AND INTERACTIONS

One way to promote the use of formative assessment in the classroom, is through changing the activities, actions and interactions of the teachers and students in the classroom. As indicated in the model in chapter 5, doing formative assessment means eliciting, interpreting and taking action as a part of planned formative assessment and noticing, recognising and responding as a part of interactive formative assessment. To do these actions often requires teachers and students to be engaged in different teaching and learning activities. It requires that the teaching and learning activities are planned and organised to provide the opportunities for the teacher and students to do formative assessment. A lecture-type of teaching activity for 60 minutes does not allow many opportunities for formative assessment. Small discussion groups or small groups doing investigative work provide many opportunities for the teacher and students to undertake formative assessment. Online teaching, using discussion groups in a 'chat-room' or 'class forum' format allows for more opportunities yet again, for doing formative assessment.

Doing formative assessment also means encouraging specific kinds of interaction in the classroom, in which feedback and feedforward is given. We use the term 'feedback' to refer to a response during formative assessment on information elicited about the student's learning. For example, in reply to a student's response, a teacher may say 'yes, a bee is an animal'. Feedback is given to the student about the correctness of their learning, that is, whether they have reached the desired learning goal. We use the term 'feedforward' (Beckett, 1996) to refer to a response that indicates what a student might do in addition, to close the gap between what they know and what is required of them. For example, the teacher may also add 'a bee is an animal as it is a consumer'. The feedforward is indicating to the student that they must also know the criteria on which we decide whether something is an animal or not.

But to do these activities, actions and interactions, other changes must also occur in the classroom. For many teachers (and students), it also means a change in the way they think about teaching and learning, and about teachers and learners. This may require a change in the teachers' conceptions about the learning process and their role as a teacher in it, especially the integral nature of teaching, learning and formative assessment. It may require accepting that both teachers and students will have a perspective on the formative assessment process and that these may differ. It may require the development of a new understanding of the proactive role of students in formative assessment. It may require knowledge of how to teach the students to take a more active role in their formative assessment.

To do formative assessment, teachers also need to be confident in their content knowledge, their knowledge of the students in the class, their knowledge of common alternative conceptions held by students, their knowledge of the progression of understanding of students in a particular curriculum area, and their knowledge of how to develop the student's conceptions. Knowing these will give a teacher more confidence to undertake formative assessment.

Hence, when a student says 'there is no gravity on the moon because there is no atmosphere', a teacher needs several types of knowledge to undertake the formative assessment. Firstly, she or he is helped if they know the scientific understanding of gravity on the moon to be able to judge that this student's comment is scientifically unacceptable. Secondly, she or he is helped if they know that this is a commonly held alternative conception and therefore other students in the class will probably also hold this idea. Thirdly, she or he is helped if they know of a learning activity that will mediate the students' learning towards the scientifically acceptable ideas of gravity. In other words, a teacher's knowledge of and confidence to do formative assessment depends on their content knowledge, pedagogical knowledge and pedagogical content knowledge.

Another aspect of a teacher's confidence is her or his disposition or orientation to uncertainty. Changing one's teaching and doing formative assessment, both require teachers to tolerate uncertainty, to be flexible and to take risks. For example, if a teacher does find out in the busyness of a lesson, that half of the students already know the concepts to be learnt in the lesson, what is he or she to do? Being responsive as a part of formative assessment involves the teacher in risk, uncertainty and flexibility. This may be an uncomfortable position for a teacher who is inexperienced or under stress or lacking in confidence.

We feel that an important prerequisite to doing formative assessment in the classroom is the establishment of a relationship of trust between the teacher and students. With trust, the disclosure by the students will be greater and more helpful for giving feedback and feedforward to teachers and students. As indicated in chapter 4, students take a risk when disclosing what they know and can do (or do not know or cannot do) to the teacher. They risk having the gaps in their learning exposed to other students, recorded and reported to others if the information elicited is also used for summative purposes, being ridiculed or put-down, or feeling uncomfortable and not-cool when they get feedback about their non-learning. The degree to which students disclose is mediated by cognitive, affective, social and relational factors. Developing a sense of trust in the classroom is crucial before any significant degree of formative assessment can be done.

Other changes required for the undertaking of formative assessment in the classroom include the enabling of students to be not just involved but proactive in formative assessment; the negotiation of assessment criteria between teacher and students; and the willingness of the teacher (and students) to be responsive to formative assessment information.

To make these changes in the classroom, professional development and curriculum development may be required. These are now discussed.

8.2   CURRICULUM DEVELOPMENT

Curriculum development is another way to promote the practice of formative assessment in the classroom. By 'curriculum development' is meant the development of curricula, both nationally (or state-wide) and in the classroom. Curricula that promote the practice of formative assessment are those that enable those teaching and learning activities, during which the teacher can do formative assessment if appropriate. Such curricula are also those that enable the planning for specific formative assessment activities, such as a brainstorm; that allow for flexibility for the teacher to make teaching decisions is the busyness of the classroom; that allow students to negotiate the criteria upon which they will be formatively assessed and given feedback on their learning; and that allow students to be proactive in being involved in formative assessment. A model of curriculum development that is useful here is that by McGee (1997) as it is based on the notion of curriculum development as teacher decision making.

8.3   PROFESSIONAL DEVELOPMENT

The professional development of the teachers was also monitored during the research documented in this book. As part of the research project, eleven teacher development days were held over the two years of the project (see Bell and Cowie, 1997, appendix 1). These teacher development days were included in the research design so that the ten teachers and two researchers could meet:

• to reflect on past and to reflect for future assessment practices in science classrooms

• for the input of new ideas for assessment in science classrooms from each other or from guest speakers

• to discuss the trialing of new assessment activities in their classroom in between meetings

• to discuss the data analyses and emerging model of formative assessment.

These four activities had been shown to promote teacher development (Bell and Gilbert, 1996) and the format of the eleven meetings was based on these research findings. Details of the actual teacher development activities are documented in Bell and Cowie, 1997, pp 260-261

The teachers indicated that several key activities had facilitated their professional development on formative assessment and they are reported here as possible activities for other teacher educators to use.

Discussing concerns about assessment in general.

In the discussions on the teacher development days, the teachers were able to raise and discuss their concerns about assessment in science, including formative assessment. In 1995, the teachers' main concern was how formative assessment interfaced with the other assessment developments and issues in schools at the time. This acknowledges that we cannot consider formative assessment in isolation from other purposes of assessment or from national, state and school policies on assessment. Formative assessment is integrated and highly contextualised. At their request, speakers from the Ministry of Education and the Education Review Office were invited to clarify, for the teachers, the policies and developments in assessment, which affected what they as teachers did in the classroom. The main concerns about assessment (in general) were recording and reporting, especially to parents; assessments used for the purposes of the Education Review Office's auditing and accountability reviews; some of the Ministry of Education's policies on assessment; and assessment of learning with reference to the (then) new science curriculum. In discussing these concerns, the teachers were able to clarify the multiple purposes for assessment, including the purposes for formative assessment.

Discussing the nature and purpose of formative assessment.

In the second year of the research project, the discussions in the teacher development days tended to be more focussed on formative assessment, once their concerns about assessment in general had been acknowledged and addressed. The main concerns raised were that of the confusion about what formative assessment is; the difficulty in defining formative assessment; and the different meaning for formative assessment depending whether the teacher is checking to see if the intended learning has been learnt, or what actual learning has occurred (Bell and Cowie, 1997, pp263-265)

The concern of the different purposes for doing formative assessment for teachers and students is also pertinent here. For many teachers, the manageability of assessment in the classroom may mean that they collect data that can be used for both formative and summative purposes. But for students, using the same data for different purposes (formative and summative), places them in a situation of risk, as indicated in cameo 6.4 Density and the Tower. Do they disclose their confusion or lack of understanding if the formative assessment information is also to be for summative purposes?

Teachers' knowledge

In the course of the eleven teacher development days, many topics connected with formative assessment were discussed. As a result of these debates, the professional knowledge of the teachers was developed. We feel that the most important of these was that of views of learning and the role of the teacher in mediating the students' learning.

Making the tacit, explicit

Whilst the teachers were already doing formative assessment, they mentioned that it was largely a tacit part of their teaching. They valued the opportunities to develop a language to talk with each other about formative assessment and to articulate what it was they were doing that could be called formative assessment. In doing so, the teachers shared activities that they had done in their classrooms which might be called 'formative assessment'. Whilst, this sharing of classroom activities helped clarify what was meant by formative assessment, the sharing of professional ideas was also a valued part of the professional development (Bell and Cowie, 1997, pp 266-267). The activities shared included specific formative assessment activities; learning activities that created opportunities for the teacher to carry out formative assessment; and ways to introduce flexibility into the school scheme. The ideas shared also included ways to explain formative, self- and peer assessment to students and in particular, the students' role in formative assessment.

Reflection and anticipation

As a part of their professional development, the teachers did both reflection on teaching and anticipation of what they might do differently. Like Beckett (1996), we see teachers needing to do both reflective and anticipative action in the process of changing what they did in the classroom. Being able to visualise and conceptualise the new teaching and assessment actions helped the teachers to do the new practices in the classroom. Whilst the model (in chapter 5) helped the teachers to reflect on their actions in the classroom, the anticipation was helped by the sharing of their classroom experiences.

The sharing of concerns and problems with doing formative assessment in the classroom.

The sharing times in the teacher development days helped the teachers to collectively discuss what they were doing that was formative assessment, and to find solutions to the difficulties they were having in doing formative assessment in the classroom. The concerns raised during these discussions included the difficulty in planning for formative assessment as planned formative assessments do not always fit into the flow of the lesson; the unstructured nature of formative assessment; the difficulty in assessing each student in-depth each lesson; the problem of too much information being collected; the length of time needed to record anecdotal notes for formative assessment; the time needed to consider the formative assessment information obtained; the need to have a range of formative assessment strategies; the problem that formative assessment is not able to be done effectively if the teacher is under stress or tired; the concerns that the demands to increase summative assessments may decrease the responsiveness of formative assessments; the recording of formative assessments – is it needed?; the need for students to know about formative assessment and the need for students to be taught self- and peer-assessment (Bell and Cowie, 1997, pp 263-265). In summary, the main concerns about doing formative assessment were the difficulty in planning and preparing for formative assessment; the need for a greater range of suggested formative assessment activities to use in the classroom; the demands on the teacher of the interactive and responsive nature of formative assessment; and the issues surrounding the recording and reporting of formative assessment information. Giving time for the discussion of these concerns was an important part of the professional development of the teachers.

Feedback on the changes in their teaching

The teachers also valued receiving feedback on the increased students' learning as a result of their doing formative assessment. Teacher development is helped if there is perceived value in the new practice or knowledge (Bell and Gilbert, 1996, p. 70). The teachers indicated that the main value of formative assessment to them was to find out what learning and thinking was occurring during the learning episodes; to monitor progress by students and teachers; to aid the planning and re-planning of their teaching; and to provide qualitative information to supplement the quantitative marks on achievement reported to students, caregivers and the school. The teachers indicated that their main uses of formative assessment were to obtain information that indicated what learning had occurred; information that could be used in planning and evaluating of the teaching and learning in their classrooms; and information about the teaching, learning and assessment processes themselves (Bell and Cowie, 1997, pp274-275).

Other professional development activities

The teachers also indicated that other activities helped their understanding of and use of formative assessment in their classrooms. These were learning about the research findings, having a colleague or the researcher to discuss classroom events with, the reading material handed out in the project, the input of the facilitators, listening to others in the group, the guest speakers, and learning about assessment activities from teachers in another level of schooling.

Activities outside of the project were also mentioned: personal reading, being involved in writing a school policy on assessment or the development of assessment activities for the school, other curriculum development courses, performance appraisal workshops and implementation, strategic planning workshops and implementation, quality management work, and visiting and observing in other schools (Bell and Cowie, 1997, pp267-269).

In the second to last teacher development day, the teachers were asked for their recommendations for activities for future teacher development workshops that might be run for other teachers on formative assessment. The main recommendation was activities that would raise teachers' awareness of the nature of formative assessment and what formative assessment they were currently doing. Such activities would include:

• planned formative assessment activities for the teachers to try in their classrooms and discuss as a group
  • observers (the facilitator or another teacher) in the teachers' classroom
  • videotaping lessons for group discussion
  • discussing video clips to illustrate parts of the model of formative assessment
  • discussing transcripts or other data (for example, the cameos) from the research
  • discussions on the importance of the interactive formative assessment
  • discussion with students about formative assessment in the workshops
  • reflection on the teachers' own and other teachers' practice
(Bell and Cowie, 1997, pp269-270).

The teachers in the research project suggested that planned formative assessment be first addressed in the workshops, then interactive formative assessment, and that formative assessment be looked at in a number of curricular areas. And as already documented, the teachers found it helpful to have guest-speakers to clarify and address their concerns about assessment policy in general.

In conclusion, we wish to comment that while this research was not designed to research if formative assessment improves learning or not, we are excited by the research or others (for example, as reviewed by Black and Wiliam, 1996) that indicates that this is so. We hope that the contribution of this book is to make a largely tacit process, more explicit, so that other teachers can promote the use of formative assessment in their teaching.

APPENDIX

THE DATA CODING

All data collected and reported was coded to provide a reference to the data and to protect the anonymity of the data sources.

The case study interviews

In the case studies reported in this book:

The teacher end-of-lesson interviews are coded, for example (T2/D8/96) where T2 represents teacher 2, D 8 the discussion after lesson 8 with this teacher, in 1996.

The teacher end-of-unit interviews are coded, for example, T2/I/96, where T2 is teacher 2, I an end-of-unit interview, in 1996.

The teacher end-of-year interviews are coded, for example, T2/EOY/96, indicating the 1996 end-of-year interview with teacher 2.

The student individual interviews are coded, for example, (S24/I/95a) which is the code for an interview with student 4 of Teacher 2 in phase 1 in 1995. Individual student end-of-unit and end of lesson discussions in phase 2 in 1995 and 1996 are coded (Sxx/I/95b) and (Sxx/I/96).

Student end-of-lesson group discussions are coded, for example, (SG75/L3/96) which is the code for a discussion with group 5 of Teacher 7 after lesson 3 (L3) in 1996. The student member checks are coded for example (Sxx/MC/96).

The classroom observations

The field notes are coded, for example (T2/FN4/96) which indicates the item is a field note (FN), taken during lesson four of teacher 1, in 1996. The head notes and documentary data were recorded in the field notes and had no separate code.

The surveys

The quotations from the two teacher surveys, recorded in this book, are coded with an 'Su' for 'survey', the number of the survey, the year, the number of the teacher and the question number. For example, code Su1/96/2/T8 indicates that this was a response in survey 1 in 1996 to question 2, by teacher 8.

The teacher development day discussions

The analysis of the audiotaping transcripts was done by coding each distinct segment of discussion. Hence, the quotations documented may represent the speech of more than one person as it is of a segment of talking, rather than the contribution of a single person. A segment might be the ideas of one or several teachers. The code used to identify the quotations is, for example, TD10/96/28.2, referring to the second data segment, on tape 28, recorded on the tenth teacher development day which was held in the second year of the project, 1996.

Field notes were also taken at these days and used to inform the data analysis. No coded data from these field notes is recorded in this book.

REFERENCES

Augoustinos and Walker (1995) *Social Cognition: an integrated introduction.* London: Sage Publications.

Bachor, D. G, and Anderson, J. (1994) Elementary Teachers' Assessment Practices as Observed in the Province of British Columbia. *Assessment in Education* , 1(1), 63-93.

Bachor, D., Anderson, J., Walsh, J., and Muir, W. (1994) Classroom Assessment and the relationship of representativeness, accuracy and consistency. *The Alberta Journal of Educational Research* , XL(2), 247-262.

Bahktin, M. (1986) *Speech genres and other late essays.* Eds C. Emerson and M. Holquist, trans, V. McGee. Austin: University of Texas Press.

Bauersfeld, A. (1988) Interaction, Construction and Knowledge: Alternative perspectives for mathematics education. In D. Grouws and T. Cooney (1988) *Perspectives on research in effective mathematics teaching.* Virginia: Lawrence Erlbaum Press.

Beckett, D. (1996) Critical Judgement and Professional Practice. *Educational Theory,* 46, 2, 135-149.

Bell, B. (1993a) *Taking into account students' thinking: a teacher development guide.* Hamilton: University of Waikato.

Bell, B. (Ed.). (1993b) *I Know about LISP But How do I Put it into Practice?* Final Report of the Learning in Science Project (Teacher Development). Hamilton: University of Waikato.

Bell, B. (1995) Interviewing: a technique for assessing science knowledge. In S. Glynn and R. Duit (1995) (Eds) *Learning Science in Schools: research reforming practice.* Mahwah, New Jersey: Lawrence Erlbaum Associates, Publishers.

Bell, B. (in press) Formative assessment and science education; a model and theorising. In R. Millar, J. Leach, J. Osborne (Eds) (in press*) Improving Science Education: the Contribution of Research.* Buckingham, Open University Press.

Bell, B. and Cowie, B. (1997) *Formative Assessment and Science Education.* Research Report of the Learning in Science Project (Assessment), August, 1997. Hamilton: University of Waikato, pp. 340.

Bell, B. and Cowie, B. (1999) Researching Formative Assessment. In J. Loughran (Ed) (1999) *Researching teaching: methodologies and practices for understanding pedagogy.* London: Falmer Press.

Bell, B. and Gilbert, J. (1996) *Teacher development: a model from science education.* London: Falmer Press.

Bell, B. and Pearson, J. (1992) 'Better Learning'. *International Journal of Science Education,* 14 (3) 349-361.

Bennett, N., Desforges, C., Cockburn, A., and Wilkinson, B. (1984) *The Quality of Pupil Learning Experiences.* London: Lawrence Erlbaum Associates.

Berlak, H. (1992a) The Need for New Science of Assessment. In H. Berlak, F. M. Newmann, E. Adams, D. A. Archbald, T. Burgess, J. Raven, and T. A. Romberg (Eds.), *Toward a New Science of Educational Testing and Assessment* (pp. 1-22). Albany: State University of New York Press.

Berlak, H. (1992b) Toward the Development of a New Science of Educational Testing and Assessment. In H. Berlak, F. M. Newmann, E. Adams, D. A. Archbald, T. Burgess, J. Raven, and T. A. Romberg (Eds.), *Toward a New Science of Educational Testing and Assessment* Albany: State University of New York Press.

Biggs, J. (1995) Assessing for Learning: Some Dimensions Underlying New Approaches to Educational Assessment. *The Alberta Journal of Educational Research* , 41(1), 1-17.

Black, P. (1993) Formative and Summative Assessment by Teachers'. *Studies in Science Education* , 21, 49-97

Black, P. (1995a) Can teachers use assessment to improve learning? *British Journal of Curriculum and Assessment* , 5(2), 7-11.

Black, P. (1995b) Lessons in Evolving Good Practice. In B. Fairbrother and P. Black (Eds.) *Teachers Assessing Pupils.* London: The Association for Science Education.

Black, P. (1998) Assessment by teachers and the improvement of students' learning. . In Fraser, B. and Tobin, K. (1998) (Eds) *International Handbook of Science Education*. Dordrecht: Kluwer Academic Publishers

Black, P. and Wiliam, D. (1998) Assessment and Classroom Learning. *Assessment in Education*, 5 (1) 7-74.

Blackmore, J. (1988) *Assessment and Accountability*. Geelong: Deakin University Press.

Boud, D. (1995) Assessment and Learning: Contradictory or Complementary. In P. Knight (Ed.), *Assessment for Learning in Higher Education*. Birmingham: Kogan Page

Brown, S. (1996) *Summary Comment*. Symposium on Validity in Educational Assessment. Educational Assessment Research Unit, University of Otago, 28-30 June.

Brown, S. and Knight, P. (1994). *Assessing Learners in Higher Education*. London: Kogan Page.

Bruner, J. (1986). *Actual Minds, Possible Worlds*. Cambridge, Massachusetts: Harvard University Press.

Bruner, J. (1990) *Acts of Meaning*. Cambridge, MA: Harvard University Press.

Carr, M. (1998) *Early Childhood Technology Education*. Unpublished DPhil thesis, University of Waikato, Hamilton, New Zealand.

Clarke, D. (1995) Constructive Assessment: Mathematics and the Student. In A. Richardson (Ed.) *Flair: AAMT Proceedings*. Adelaide: AAMT.

Claxton, G. (1995) What Kind of Learning Does Self-Assessment Drive? Developing a 'nose' for quality: Comments on Klenowshi. *Assessment in Education* , 2(3), 333-343

Cobb, P. (1994) Where is the Mind? *Educational Researcher*, 23 (7)13-20.

Cowie, B. (1997) Formative Assessment and Science Classrooms. In B. Bell and R. Baker (Eds) (1997) *Developing the Science Curriculum in Aotearoa New Zealand*. Auckland: Longman Addison Wesley.

Cowie, B. (2000) *Formative Assessment in Science Classrooms*. Unpublished DPhil Thesis, Hamilton: University of Waikato.

Cowie, B. and Bell, B. (1995) *Learning in Science Project (Assessment) Research Report 1: Views of Assessment*. Hamilton: University of Waikato.

Cowie, B. and Bell, B. (1996) Validity and Formative Assessment in the Science Classroom. Invited Keynote Paper to *Symposium on Validity in Educational Assessment*, 28-30 June, Dunedin, New Zealand.

Cowie, B. and Bell, B. (1999) A Model of Formative Assessment in Science Education, *Assessment in Education*, 6 (1) 101-116.

Cowie, B., Boulter, C., and Bell, B. (1996). *Developing a Framework for Assessment of Science in the Classroom*. Working paper of the learning in Science project (Assessment). Hamilton: University of Waikato.

Crooks, T. J. (1988) The Impact of Classroom Evaluation Practices on Students. *Review of Educational Research* , 58(14), 438-481.

Dassa, C., Vazquez-Abad, J., and Ajar, D. (1993) Formative Assessment in a Classroom Setting: From Practice to Computer Innovations. *The Alberta Journal of Educational Research* , 39(1) 111-126.

Denscombe, M. (1995) Teachers as an Audience for Research: the acceptability of ethnographic approaches to classroom research. *Teachers and Teaching: theory and practice* , 1(2) 173-192.

Drewery, W. and Winslade, J. (1997) The Theoretical Story of Narrative Therapy. In G. Monk, G., J. Winslade, K. Crockett, and D. Epston, (Eds) (1997) *Narrative Therapy in Practice: the archaeology of hope*. San Francisco: Jossey-Bass.

Department of Education, (1989) *Assessment For Better Learning: A Public Discussion Document*. Wellington: Department of Education.

Donaldson, M. (1978) *Children's Minds*. London: Fontana.

Driver, R. (1989) Students' Conceptions and the Learning of Science. *International Journal of Science Education*, 11 (5) 481-490.

Driver, R. and Bell, B. (1986) Students' Thinking and the Learning of Science. *School Science Review*, 67 (240) 443-456.

Driver, R. and Newton, P. (1997) *Establishing the norms of scientific argumentation in classrooms*. Paper given to the European Science Education Research Association Conference, 2-6 September, 1997, Rome.

Driver, R., Asoko, H., Leach, J., Mortimer, E. and Scott, P. (1994) Constructing Scientific Knowledge in the Classroom. *Educational Researcher*, 23 (7) 5-12. (1994)

Duit, R. (1994) Research on Children's Conceptions- developments and trends. In Pfundt, H. and Duit, R. (1994) *Bibliography: Students' Alternative Frameworks and Science Education*. Kiel: IPN.

Duschl, R. and Gitomer, D. (1997) Strategies and challenges to changing the focus of assessment and instruction in science classrooms. *Educational Assessment*, 4, 37-73.

Dweck, C. (1986) Motivational processes affecting learning. *American Psychologist* (Special issue: Psychological science and education), 41, 1040-1048.

Dweck, C.S. (1989) Motivation. In A. Lesgold and R. Glaser (Eds) *Foundations for a Psychology of Education*. Hillsdale, New Jersey: Lawrence Erlbaum Associates Inc.

Education Review Office (http://www.ero.govt.nz).

Edwards, D. (1991) Categories are for talking: On the cognitive and discursive bases of categorisation. *Theory and Psychology*, 1: 515-542.

Edwards, D. and Mercer, N. (1987) *Common Knowledge the development of understanding in the classroom*. London: Methuen.

Erickson, G. and Meyer, K. (1998) Performance assessment tasks in science: what are they measuring? In B. Fraser and K. Tobin (1998) (Eds) *International Handbook of Science Education*. Dordrecht: Kluwer Academic Publishers.

Fairbrother, B. (1995) Pupils as Learners. In B. Fairbrother, P. Black, and P. Gill (Eds.), *Teachers Assessing Pupils*. London: The Association of Science Education.

Fairbrother, B., Black, P. and Gill, P. (Eds.) (1995) *Teachers Assessing Pupils*. London: The Association of Science Education.

Falchikov, N. (1995).Improving Feedback To and From Students. In P. Knight (Ed.), *Assessment for Learning in Higher Education* Birmingham: Kogan Page.

Filer, A. (1993) The Assessment of Classroom Language: Challenging the rhetoric of 'objectivity'. *International Studies in Sociology of Education* , 3(2), 193-212.

Filer, A. (1995) Teacher Assessment: Social process and social product. *Assessment in Education* , 2(1) 23-38.

Gardner, H. (1985) *Frames of Mind: the theory of multiple intelligences*. USA: Basic Books.

Gilbert, J. (1997) *Thinking 'Other-wise': Re-thinking the Problem of Girls and Science Education in the Post-Modern*. Unpublished DPhil thesis, University of Waikato, Hamilton, New Zealand.

Gipps, C. (1994) *Beyond Testing: Towards a Theory of Educational Assessment*. London: The Falmer Press.

Gipps, C. (1999) Socio-cultural Aspects of Assessment. *Review of Research in Education*, 24,

Gipps, C. and James, M. (1998) Broadening the basis of assessment to prevent the narrowing of learning. *The Curriculum Journal*, 9 (3) 285-297.

Gitomer, D. and Duschl, R. (1995) Moving towards a portfolio culture in science education. In S. Glynn and R. Duit (1995) (Eds) *Learning Science in Schools: research reforming practice*. Mahwah, New Jersey: Lawrence Erlbaum Associates, Publishers.

Gitomer, D. and Duschl, R. (1998) Emerging issues and practices in science assessment. In B. Fraser and K. Tobin (1998) (Eds) *International Handbook of Science Education*. Dordrecht: Kluwer Academic Publishers.

Glaser, R. (1963) Instructional Technology and the Measurement of Learning Outcomes: Some Questions. *American Psychologist* , 18, 519-521.

Goodfellow, J. (1996). *Weaving Webs of Caring Relationships*. Paper given to the Weaving Webs Conference: Collaborative teaching and learning in the early years curriculum, Melbourne, 11-13 July, 1996.

Gore, J. (1998) Disciplining bodies: on the continuity of power relations in pedagogy. In T. Popkewitz and M. Brennan (Eds) *Foucault's challenge: discourse, knowledge and power in education*. New York: Teachers College Press: 231-251.

Hanson, F. A. (1993) *Testing Testing : Social Consequences of an Examined Life*. Berkley: University of California Press.

Hanks, W. (1991) Foreword. In Lave, J. and Wenger, E. (1991) *Situated learning: legitimate peripheral participation*. New York: Cambridge University Press.

Hargreaves, A. (1989) *Curriculum and Assessment Reform*. Milton Keynes: Open University Press.

Harlen, W. (1995).To the Rescue of Formative Assessment. *Primary Science Review* , 37, 14-16.

Harlen, W. and James, M. (1996) *Creating a Positive Impact of Assessment on Learning*. Paper presented to the American Educational Research Association Annual Conference, New York.

Harré, R. and Gillett, G. (1994) *The Discursive Mind*. London, Sage.

Hennessy, S. (1993) Situated cognition and cognitive apprenticeship: implications for classroom learning. *Studies in Science Education*, 22, 1-41.

Hill, M. (1999) Assessment in self-managing schools: primary teachers balancing learning and accountability demands in the 1990s. *New Zealand Journal of Educational Studies*, 34 (1) 176-185.

Jaworski, B. (1994) *Investigating Mathematics Teaching : A Constructivist Enquiry*. London: The Falmer Press.

Johnson, S. (1989) *National Assessment: the APU approach*. London: Her Majesty's Stationery Office

Keeves, J. and Alagumalai, S. (1998) Advances in Measurement in Science Education. In B. Fraser and K. Tobin (Eds) *International Handbook of Science Education*. Great Britain: Kluwer Academic Publishers.

Klenowski, V. (1995) Student Self-evaluation Processes in Student-centred Teaching and Learning Contexts of Australia and England. *Assessment in Education* , 2(2), 145-165.

Kluger, A. and deNisi, A. (1996) The Effects of Feedback interventions on Performance:a historical review, a meta-analysis, and a preliminary feedback intervention theory. *Psychological Bulletin*, 119, 254-284.

Lave, J. and Wenger, E. (1991) *Situated Learning: Legitimate Peripheral Participation*. Cambridge: Cambridge University Press.

Learvitt, R. L. (1994) The emotional culture of infant toddler day care. In J. A. Hatch (Ed) (1994) *Qualitative research in early childhood settings*. London: Praeger.

Lemke, J. (1990) *Talking Science: language, learning, values*. Norwood, NJ: Ablex.

Lepper, M. and Hodell, M. (1985) Intrinsic motivation in the classroom. In C. and A. Ames (Eds) *Research in motivation in education, vol 3: Goals and cognitions*. New York: Academic Press.

McGee, C. (1997) *Teachers and Curriculum Decision-making*. Palmerston North, New Zealand: Dunmore Press.

Mehan, H. (1979) *Learning Lessons: social organisation in the classroom*. Cambridge, Massachusetts: Harvard University Press.

Meltzer, L. and Reid, D. (1994) New Directions in the Assessment of Students with Special Needs: The Shift Toward a Constructivist Perspective. *The Journal of Special Education* , 28 (3), 338-355.

Messick, S. (1989) Validity. In R. Linn (Ed.), *Educational Measurement* Washington: Macmillan.

Ministry of Education, (1990) *Tomorrow's Standards*. Wellington: Learning Media.

Ministry of Education, (1993a) *The New Zealand Curriculum Framework*. Wellington: Learning Media.

Ministry of Education, (1993b) *Science in the New Zealand Curriculum*. Wellington: Learning Media.

Ministry of Education, (1994) *Assessment: Policy to Practice*. Wellington: Learning Media.

Monk, G., Winslade, J., Crockett, K. and Epston, D. (Eds) (1997) *Narrative Therapy in Practice: the archaeology of hope*. San Francisco: Jossey-Bass.

Moss, P. A. (1994). Can There be Validity Without Reliability? *Educational Researcher* , 23(2), 5-12.

National Education Monitoring Project, New Zealand, (http://nemp.otago.ac.nz)

National Research Council (1999) *The Assessment of Science Meets the Science of Assessment*. Board on Testing and Assessment Commission on Behavioural and Social Sciences and Education, National Research Council. Washington, DC: National Academy Press.

Newman, D., Griffin, P., and Cole, M. (1989) *The Construction Zone: Working for Cognitive Change in School*. Cambridge: Cambridge University Press.

Nuthall, G. (1997) Understanding Student Thinking and Learning in the Classroom. In B. J. Biddle, T.C. Good and I. Goodson (Eds) (1997) *The International Handbook of Teachers and Teaching*. Dordrecht: Kluwer Academic Publishers.

O'Loughlin, M. (1992) Rethinking Science Education: Beyond Piagetian constructivism toward a sociocultural model of teaching and learning. *Journal of Research in Science Teaching*, 29, 791-820.

Osborne, J. (1996) Beyond Constructivism. *Science Education*, 80 (1) 53-82.

Osborne, R. J., and Freyberg, P. S. (1985). *Learning in Science: the implications of children's science.* Auckland: Heinemann.

Osborne, R. and Wittrock, M. (1985) The Generative Learning Model and its Implications for Science Education. *Studies in Science Education*, 12, 59-87.

Parkin, C. and Richards, N. (1995). Introducing Formative Assessment at KS3: an attempt using pupil self-assessment. In B. Fairbrother, P. Black and P. Gill (Eds.), *Teachers Assessing Pupils.* London: The Association for Science Education.

Pea, R. (1993) Distributed intelligence and designs for education. In G. Salomon (Ed) (1993) *Distributed Cognitions: Psychological and educational considerations.* New York: Cambridge University Press.

Perkins, D. (1993) Person-plus: a distributed view of thinking and learning. In G. Salomon (Ed) (1993) *Distributed Cognitions: Psychological and educational considerations.* New York: Cambridge University Press.

Perrenoud, P. (1991) Towards a Pragmatic Approach to Formative Evaluation. In P. Weston (Ed.) (1991) *Assessment of Pupil Achievement : Motivation and School Success* Amsterdam: Swets and Zeitlinger.

Perrenoud, P. (1998) From Formative Evaluation to a Controlled Regulation of Learning Processes. Towards a Wider Conceptual field. *Assessment in Education*, 5 (1) 85-102.

Peterson, P. L., and Clark, C. M. (1978) Teachers' reports of their cognitive processes during teaching. *American Educational Research Journal*, 15 (4) 555-566.

Pintrich,P., Marx,R. and Boyle, R. (1993) Beyond Cold Conceptual Change: the role of motivational beliefs and classroom contextual factors in the process of conceptual change. *Review of Educational Research*, 63, 167-199.

Pryor, J. and Torrance, H. (1996) Teacher-pupil Interaction in Formative Assessment: Assessing the work or protecting the child? *The Curriculum Journal*, 7, 205-226.

Radnor, H. (1994) The problems of facilitating qualitative formative assessment in pupils. *British Journal of Educational Psychology* , 64, 145- 160.

Ramaprasad, A. (1983) On the Definition of Feedback. *Behavioural Science* , 28(1), 4-13.

Raven, J. (1992) A Model for Competence, Motivation, and Behaviour, and a Paradigm for Assessment. In H. Berlak, F. M. Newmann, E. Adams, D. A. Archbald, T. Burgess, J. Raven and T, Romberg (Eds) *Towards a New Science of Educational Testing and Assessment.* Albany: State University of New York Press.

Resnick, L. (1991) Shared Cognition: Thinking a Social Practice. In L. Resnick, J. Levine, and S. Teasley (Eds) (1991) *Perspectives on Socially Shared Cognition.* Washington, DC: American Psychological Association.

Rogoff, B. (1993) Children's Guided Participation and Participatory appropriation of sociocultural activity. In R. Wozniak and K. Fischer (Eds) *Development in context: acting and thinking in specific environments.* Hillsdale, NJ: Lawrence Erlbaum Associates.

Rogoff, B. (1995) Observing sociocultural activity on three planes. In J. Wertsch, P. Del Río, A. Alvarez. (1995) (Eds) *Sociocultural Studies of Mind.* Cambridge: Cambridge University Press.

Rowe, M. B. (1987) Wait Time: Slowing Down May Be a Way of Speeding Up. *American Educator* , 11(1), 33-43,47.

Sadler, R. (1989) Formative Assessment and the Design of Instructional Systems. *Instructional Science* , 18(2) 119-144.

Sadler, R. (1998) Formative assessment: revisting the territory. *Assessment in Education*, 5 (1) 77-84.

Salomon, G. (1993a) Editors Introduction. In G. Salomon (1993) *Distributed cognitions: psychological and educational considerations.* New York: Cambridge University Press.

Salomon, G. (1993b) No distribution without individuals' cognition. In G. Salomon (1993) *Distributed cognitions: psychological and educational considerations.* New York: Cambridge University Press.

Salomon, G. and Perkins, D. (1998) Individual and social aspects of learning. *Review of Research in Education*, 23, pp 1-24.

Savage, J. and Desforges, C. (1995) The Role of Informal Assessment in Teachers' Practical Action. *Educational Studies* , 21(3), 433-446.

Scott, P. (1997) *Developing science concepts in secondary classrooms: an analysis of pedagogical interactions from a Vygotskian perspective.* Unpublished PhD thesis, University of Leeds, UK.

Scott, P. (1998) Teacher talk and meaning making in science classrooms: a Vygotskian analysis and review. *Studies in Science Education*, 32, 45-80.

Scott, P. (1999) *An analysis of science classroom talk in terms of the authoritative and dialogic nature of the discourse*. Paper presented to the 1999 NARST Annual Meeting, Boston, USA.

Shulman, L. (1987) Knowledge and teaching: foundations of the new reforms. *Harvard Educational Review*, 57, 1-22.

Stiggins, R. J. (1991) Assessment Literacy for the 21st Century. *Phi Delta Kappan* , 77(3).

Sutton, R. (1995) *Assessment and Learning*. Auckland: Auckland College Printery.

Tamir, P. (1998) Assessment and evaluation in science education: opportunities to learn and outcomes. In B. Fraser and K. Tobin (1998) (Eds) *International Handbook of Science Education*. Dordrecht: Kluwer Academic Publishers.

Tasker, R. amd Osborne, R. (1985) Science Teaching and Science Learning. In R. Osborne and P. Freyberg (1985) *Learning in Science: the implications of children's science*. Auckland: Heinemann.

Torrance, H. (1993) Formative Assessment : Some Theoretical Problems and Empirical Questions. *Cambridge Journal of Education* , 23(3), 333-343.

Torrance, H., and Pryor, J. (1995). Investigating Teacher Assessment in Infant Classrooms: methodological problems and emerging issues. *Assessment in Education* , 2(3), 305-320.

Tunstall, P., and Gipps, C. (1995). *How does your teacher help you to make your work better? Children's Understanding of Formative Assessment*. Paper presented to the annual conference of the British Educational Research Association, Bath, England.

Vygotsky, L. (1978*) Mind in Society: the development of higher psychological processes*. Edited by Cole, M., John-Steiner, V., Scribner, S. and Souberman. Cambridge, MA: Harvard University Press.

Weinstein, R. S. (1989). Perceptions of Classroom Processes and Student Motivation: Children's Views of Self-Fulfilling Prophecies. In C. A. and. A. Ames (Ed.), *Research on Motivation in Education Volume 3 : Goals and Cognitions* New York: Academic Press.

Wertsch, J. (1991) *Voices of the Mind*. Cambridge, Massachusetts, Harvard University Press.

Wertsch, J., Del Río, P., Alvarez, A. (1995) Sociocultural studies: history, action and mediation In J. Wertsch, P. Del Río, A. Alvarez (Eds) *Sociocultural Studies of Mind*. Cambridge: Cambridge University Press.

Wertsch, J., Del Río, P., Alvarez, A. (1995) (Eds) *Sociocultural Studies of Mind*. Cambridge University Press.

Wetherell, M. and Potter, J. (1992) *Mapping the language of racism: Discourse and the legitimation of exploitation*. Hemel Hempstead: Harvester Wheatsheaf.

White, R., and Gunstone, R. (1992) *Probing Understanding*. New York: Falmer Press .

Wiliam, D. (1992) Some Technical issues in Assessment : a user's guide. *British Journal of Curriculum and Assessment* , 2(3) 11-20.

Wiliam, D. (1994) *Towards a Philosophy for Educational Assessment*. Paper presented to the annual conference of the British Education Research Association, Oxford.

Wiliam, D. and Black, P. (1995). Meanings and consequences: a basis for distinguishing formative and summative functions of assessment? Paper presented at the annual conference of the British Educational Research Association, Bath, England.

Willis, D. (1994). School-Based Assessment: Underlying ideologies and their implications for teachers and learners. *New Zealand Journal of Educational Studies* , 29(2), 161-174.

INDEX

Science & Technology Education Library

Series editor: Ken Tobin, *University of Pennsylvania, Philadelphia, USA*

Publications

1. W.-M. Roth: *Authentic School Science*. Knowing and Learning in Open-Inquiry Science Laboratories. 1995   ISBN 0-7923-3088-9; Pb: 0-7923-3307-1
2. L.H. Parker, L.J. Rennie and B.J. Fraser (eds.): *Gender, Science and Mathematics*. Shortening the Shadow. 1996   ISBN 0-7923-3535-X; Pb: 0-7923-3582-1
3. W.-M. Roth: *Designing Communities*. 1997
          ISBN 0-7923-4703-X; Pb: 0-7923-4704-8
4. W.W. Cobern (ed.): *Socio-Cultural Perspectives on Science Education*. An International Dialogue. 1998   ISBN 0-7923-4987-3; Pb: 0-7923-4988-1
5. W.F. McComas (ed.): *The Nature of Science in Science Education*. Rationales and Strategies. 1998   ISBN 0-7923-5080-4
6. J. Gess-Newsome and N.C. Lederman (eds.): *Examining Pedagogical Content Knowledge*. The Construct and its Implications for Science Education. 1999
          ISBN 0-7923-5903-8
7. J. Wallace and W. Louden: *Teacher's Learning*. Stories of Science Education. 2000
          ISBN 0-7923-6259-4; Pb: 0-7923-6260-8
8. D. Shorrocks-Taylor and E.W. Jenkins (eds.): *Learning from Others*. International Comparisons in Education. 2000   ISBN 0-7923-6343-4
9. W.W. Cobern: *Everyday Thoughts about Nature*. A Worldview Investigation of Important Concepts Students Use to Make Sense of Nature with Specific Attention to Science. 2000   ISBN 0-7923-6344-2; Pb: 0-7923-6345-0
10. S.K. Abell (ed.): *Science Teacher Education*. An International Perspective. 2000
          ISBN 0-7923-6455-4
11. K.M. Fisher, J.H. Wandersee and D.E. Moody: *Mapping Biology Knowledge*. 2000
          ISBN 0-7923-6575-5
12. B. Bell and B. Cowie: *Formative Assessment and Science Education*. 2001
          ISBN 0-7923-6768-5; PB: 0-7923-6769-3

KLUWER ACADEMIC PUBLISHERS – DORDRECHT / BOSTON / LONDON